CAMBRIDGE EARTH SCIENCE SERIES

Editors:

W. B. Harland, A. H. Cook, N. F. Hughes
J. C. Sclater, S. W. Richardson

Phanerozoic paleocontinental world maps

Phanerozoic paleocontinental world maps

A. G. SMITH
Lecturer in Geology, University of Cambridge
A. M. HURLEY
Data Administrator, British National Oil Corporation
J. C. BRIDEN
Professor of Geophysics, University of Leeds

CAMBRIDGE UNIVERSITY PRESS
CAMBRIDGE
LONDON NEW YORK NEW ROCHELLE
MELBOURNE SYDNEY

CAMBRIDGE UNIVERSITY PRESS
Cambridge, New York, Melbourne, Madrid, Cape Town,
Singapore, São Paulo, Delhi, Tokyo, Mexico City

Cambridge University Press
The Edinburgh Building, Cambridge CB2 8RU, UK

Published in the United States of America by Cambridge University Press, New York

www.cambridge.org
Information on this title: www.cambridge.org/9780521232586

This is a revised and enlarged version of *Mesozoic and
Cenozoic Paleocontinental Maps* by A. G. Smith and
J. C. Briden and © Cambridge University Press 1977

© Cambridge University Press 1980

First published 1980
Re-issued 2011

A catalogue record for this publication is available from the British Library

ISBN 978-0-521-23257-9 Hardback
ISBN 978-0-521-23258-6 Paperback

Cambridge University Press has no responsibility for the persistence or
accuracy of URLs for external or third-party internet websites referred to in
this publication, and does not guarantee that any content on such websites is,
or will remain, accurate or appropriate.

Contents

Introduction

Section 1 contains maps that may be regarded as modifications and improvements of the previously published map book (Smith & Briden, 1977). Section 2 contains previously unpublished Paleozoic composite maps. The projections shown are chosen as a result of users' comments about the previous publication. We hope that the maps in this book will be useful to all teachers and research workers concerned with large-scale geological and geophysical problems. There are eighty-eight maps in all, drawn in four series at twenty-two periods of time: present day, 10 Ma, then at 20-Ma intervals back to 220 Ma, and then at 40-Ma intervals from 240 Ma back to 560 Ma. The four series consist of a cylindrical equidistant series, split along the zero meridian into two halves to produce two maps, and two polar Lambert equal-area series.

As the title shows, the emphasis is on the past relative positions of the continents. Of necessity, the oceans separate continents from one another, but no attempt has been made to show any former oceanic feature other than the approximate edges of the ocean basins. It is hoped that readers will plot their own paleogeographic, paleontologic or paleoclimatic data on these maps, or use them to make their own plate tectonic interpretations of the time concerned. The paleo-positions of oceanic plate boundaries may be estimated by referring to the original literature.

Most of us depend heavily on the recognition of coastlines to locate our positions on present-day geographic maps. Yet the coastline is one of the most ephemeral features of paleography, and only in exceptional situations can it be located in the stratigraphic record. Thus the coastlines drawn on the maps in this book are merely our estimates of the past positions of present-day coastlines. Since these features did not exist in the past, their main value is that of aiding the recognition of the past positions of continental fragments. The continental shelves at the edges of the continents are much more enduring than the coastlines. The maps show the present-day 1000 m (500 fathom) submarine contour, except around most of the Pacific basin where it has been omitted. Crosses mark the approximate positions of plate boundaries along which continents are judged to have been broken in order to produce the maps.

The past positions of the present geographic latitude—longitude grid are shown within the continental fragments at intervals of ten degrees, so that paleogeologic features of the continents may be plotted in their past, original positions. The problems of drawing the present grid in areas affected by orogeny, and of drawing former continental boundaries of areas that have subsequently been welded together, are briefly discussed below.

Superimposed on each map is a paleogeographic latitude—longitude grid drawn at thirty-degree intervals. Nevertheless, the maps cannot be regarded as paleogeographic maps since they do not show past geography. They do show our estimates of former relative continental positions. In other words, they are paleocontinental maps. Because of the manner in which the Paleozoic maps have been constructed, we prefer to call them composite maps, since they are in fact a number of maps superimposed to form one world paleocontinental map. The method of making the composites is discussed in section 2.

One of the interesting features of these maps is that they are entirely machine made. Only captions have been added to the computer drawings. Though this method of production gives maps that may not be as pleasing to look at as those drawn by hand, it has two distinct advantages: the maps are available quickly and they cost less to make. Where new data become available the maps are easily corrected. Whilst a large amount of new data has been incorporated since the previous publication, we still believe these maps to be provisional estimates based on our interpretation of the published data.

The projections

For simplicity we have chosen only two projections, one cylindrical, one azimuthal, both of which give us square maps of dimensions 18 cm × 18 cm. The whole world cylindrical projection has been split into two maps which are shown on opposite pages for each age. The left-hand map centre is at latitude 0°, longitude −90°, and the right-hand map centre is at latitude 0°, longitude 90° (latitude and longitude being counted positive northwards and eastwards respectively, and the zero of paleolongitude being, inevitably, arbitrary). The Lambert equal-area projections give a north and south polar view of the Earth. Ellipticity corrections are negligible for world maps on the scale used. In all the maps a thirty-degree latitude–longitude grid is superimposed.

Cylindrical equidistant projection

This projection is a simple $X-Y$ grid, like a sheet of graph paper, where the vertical axis (Y) is latitude and the horizontal axis (X) longitude. Lines of latitude are equally spaced parallel lines, whilst lines of longitude are orthogonal equally spaced parallel lines. Since both the height and width of the maps are 18 cm, there is a convenient scale of 1 cm = 10 degrees. The main advantages of this projection over the more familiar Mercator projection are that one can project the entire world, albeit with considerable distortion at the poles, and that it is simple to locate specific latitude–longitude positions using a ruler or centimetre graph paper.

Lambert equal-area projection

As the name indicates, this projection preserves the relative differences between the areas on the globe. In the polar case, the construction is fairly simple.

In the series shown, the maps' centre is either the north or south pole: $(90°, 0°)$ or $(−90°, 0°)$. The lines of latitude form concentric rings about the pole, whilst the lines of longitude are equally spaced radial lines. The relationship between the distance R from the map centre to a point p, and the colatitude Θ of p is

$R = K \sin \Theta/2$ cm

where $K = 12.73$. This gives a convenient scale of

1 cm$^2 = 10^{12}$ m^2.

The main advantage of this projection is its equal-area property, which is particularly useful for work involving global distributions of sediment types, faunal provinces, etc.

Section 1: Mesozoic and Cenozoic paleocontinental maps

Introduction

This section contains fifty-two maps, which cover the time period from the present day back to 220 Ma at 20-Ma intervals including four maps at 10 Ma.

Method of making the maps

The maps are made in two stages. The first stage is essentially the making of a continental reassembly after closure of the Atlantic, Indian and smaller oceans by the appropriate amount. The second stage projects the resultant paleo-continental reassembly as a map.

Stage 1

Motion between two continents takes place at one or more plate boundaries. At a given instant the motion along a particular plate boundary may be described as a rotation at a given rate about an axis passing through the Earth's centre. The motion through a particular time interval may be found by summing all the instantaneous motions in that interval. The sum is a finite rotation about an axis passing through the Earth's centre. The net motion between two continents separated by more than one active plate boundary in a time interval is simply the sum of the finite rotations that have taken place across each of the plate boundaries in that time. Because finite rotations are not commutative, that is, they do not add like vectors, care must be taken over the order in which the finite rotations are added together.

The finite rotations taking place across compressional plate margins (present-day subduction zones) cannot be inferred from the margins themselves or from the effects adjacent to the margins. Thus, in the absence of any other information, two continents whose relative motion involves the action of a compressional plate margin in the time interval concerned cannot be repositioned relative to each other. For example, it is difficult to determine the exact position of the Pacific plate relative to the North American continent since the only common plate boundary is compressional.

In principle, the finite rotations taking place across extensional and/or translational plate margins can be determined. In general such motions create aseismic ocean basins like most of the present-day Atlantic and Indian Oceans. Provided that adequate geophysical surveys have been made of such areas, the finite rotations necessary to describe the relative motions of the surrounding continents can readily be estimated. In most cases the rotations are obtainable by matching the corresponding pairs of ocean-floor magnetic anomalies. Earlier shapes and sizes of the presently expanding ocean basins may be estimated by 'winding back' the ocean floor by the amount the basins have grown since the time of interest.

Because most of the floor that has been formed within the Atlantic and Indian Oceans is still preserved in those oceans (Heirtzler, Dickson, Herron, Pitman & Le Pichon, 1968; McKenzie & Sclater, 1971; Pitman & Talwani, 1972), the former relative positions of the continents around those oceans may be estimated at all times since their creatio To make a continental reassembly, one of the continents is chosen as a reference and all the others are repositioned relative to it. Africa has been used as the reference continent for the present series of maps, but any other continent or continental fragment around the Atlantic or Indian Oceans could equally well have been used as a reference.

Available data are sufficiently widely distributed that all the major continental fragments may be approximately repositioned relative to one another as far back as the openin of the Atlantic (Le Pichon, Sibuet & Francheteau, 1977) and Indian Oceans (Norton & Sclater, 1979). The oldest known part is the Atlantic Ocean that lies between Africa and North America. It probably began to form in very early Jurassic time. Some parts of the Indian Ocean may be of similar age but the data are inadequate to show the exact age.

The choice of rotations differs slightly from that of Smith & Briden (1977). In particular, the majority of data for the final reconstruction of Pangea was previously based on computerised fits of the 500 fathom (1000 m) contour (Bullard, Everett & Smith, 1965; Smith & Hallam, 1970). A number of recent geophysical surveys, coupled with increased magnetic anomaly data, have shown that improved fits may be made by using the additional constraints of rift markers such as early transform faults. This affects the pre-rift position of Australia with respect to Antarctica (Norton & Molnar, 1977), North America with respect to Africa (Le Pichon et al., 1977) and South America with respect to Africa (Norton & Sclater, 1979). The fit of North America and Africa is at the 3000 m contour, to allow for the possible post-rift adjustment of the 1000 m contour.

In addition to these improvements, the Lord Howe Rise has been fixed relative to Australia using the anomaly-33 fit of Weissel & Hayes (1977). Note that the apparently large gap between New Zealand and Australia is a result of Weissel and Hayes having chosen the 3000 m contour as the continental edge: the 3000 m contour of the Lord Howe Rise extends some ten degrees to the north of New Zealand at present. Since the relative positions of the Pacific and Indian plates are not well known (Molnar, Atwater, Mammerickx & Smith, 1975), no attempt has been made to illustrate the relative motion along the Alpine fault in New Zealand. This fault is assumed to lie on the plate boundary between the Indian and Pacific plates. Apart from the above two exceptions, all other fits are based on the fit of the 1000 m (500 fathom) submarine contour. The remaining finite rotation data to make the reassemblies have been taken from the following: Ladd (1976), South America to Africa; McKenzie, Molnar & Davies (1970), Arabia to Africa; Norton & Sclater (1979), India and Antarctica to Africa; Pitman & Talwani (1972), Eurasia to North America and North America to Africa; Talwani & Eldholm (1977), Greenland to Eurasia; lastly, Weissel, Hayes & Herron (1977), Australia to Antarctica.

The three series of maps at 180, 200 and 220 Ma show one supercontinent – Wegener's Pangea. The only difference between these maps is in the orientation of Pangea relative to the paleogeographic grid.

By choosing the appropriate route it is possible to circumvent the problem of repositioning continents separated by compressional plate boundaries at any time in the past 220 Ma. For example, despite the growth of the Alpine chains, Africa may be repositioned relative to Europe by repositioning Africa relative to North America, and then North America relative to Europe. Pangea is believed to have been created by the coalescence of at least three large continents in late Paleozoic time. The positions of these continents relative to one another cannot yet be uniquely determined because all routes in the repositioning procedure run across compressional plate boundaries.

Stage 2

The second stage of map-making consists of estimating the positions of the paleogeographic poles on the reassembly and making a map projection. The best estimates of the past positions of the geographic poles are those of the mean positions of paleomagnetic poles. Most of the paleomagnetic data have been published in the compilations of Irving, McElhinny and their collaborators. These are referenced in McElhinny (1972, 1973), and McElhinny & Cowley (1977a, 1977b). Those poles satisfying certain reliability criteria (McElhinny, 1973) have been selected for making the maps.

To estimate the position of the paleogeographic grid, all the reliable north paleomagnetic poles lying within the stable parts of all the continental fragments that can be

reliably repositioned relative to one another have been examined. Those north poles whose age range lies within 10 Ma of the age of a particular reassembly are selected for the map. They are all rotated to the reference continent using the rotations that have been employed in making the reassembly. The mean paleomagnetic north pole of the rotated north poles is calculated. This mean pole is taken as the best estimate of the north geographic pole of the reassembly relative to the reference continent. Once the pole position is known, the paleogeographic latitude–longitude grid may be superimposed on the reassembly. The zero meridian of longitude is arbitrary, as in present-day maps, but all longitude differences between areas whose relative positions can be determined on the reassembly are fixed by their positions relative to the mean pole of the reassembly. The movement of the grid north pole relative to present-day Africa may be regarded as the world mean apparent polar wandering path relative to Africa.

Both stages of map-making are carried out automatically by suitable computer programs. The rotation data appropriate to each plate and each age are stored on magnetic tape along with a digitised world map broken up into suitable fragments. The map-drawing program (a modified version of R. L. Parker's SUPERMAP) will also draw maps on projections other than those presented here.

Reliability of the maps

There are two sources of error. The first lies in the construction of the reassembly and the second in the projection of the reassembly as a map. The first source includes several different kinds of errors. There are errors due to the uncertainties in the ocean-floor-spreading histories of the Atlantic and Indian Oceans; those caused by ignorance of the past positions of all continental fragments affected by orogeny; and those attributable to a lack of knowledge of the shapes and former boundaries of continental fragments that have collided with one another.

The relative positions of the continents are best known for the past 100 Ma. Prior to this time the ocean-floor-spreading anomalies are less frequently developed, less well dated and in some cases they may have been eliminated, possibly by submarine diagenetic processes.

In this and the previous publication (Smith & Briden, 1977), the Indian and South Atlantic Oceans are assumed to have begun to open at 140 Ma. There is no evidence to the contrary that the opening occurred very much later than this; in fact, it may have started very early in the Jurassic

(Norton & Sclater, 1979). However, by a simple extrapolation of ocean-floor-spreading rates, a 140-Ma date is not unreasonable.

Slightly different problems exist in the North Atlantic region, where the successive positions of Greenland relative to Eurasia have been estimated using a combination of the data of Talwani & Edholm (1977) and the final fit of Bullard *et al.* (1965). In fact, the relative movements of Greenland, Eurasia and North America are a particularly vexing problem. Available information gives grossly different answers depending on whether Greenland is reassembled to Eurasia via North America (by closing the Labrador Sea) or by directly matching anomalies in the Norway–Greenland Sea (Kristoffersen & Talwani, 1977; Talwani & Eldholm, 1977).

The positions of all those areas affected by Mesozoic and Tertiary orogenesis are unknown. The relative initial positions of the continental fragments around the Caribbean and the Mediterranean, and their evolution in time, are based mostly on speculations by Freeland & Dietz (1971) and Smith (1971). The Caribbean area has been held fixed to North America. On maps of 160 Ma and older it overlaps onto Africa; no attempt has been made to correct this overlap. Though relative motions are known to have occurred among the continental fragments bordering the Pacific basin, no attempt has been made to reconstruct these areas. The approximate outcrop areas of the Phanerozoic orogenic belts are shown on the present-day cylindrical equidistant maps (maps 1 and 2).

Prior to the collision of two or more continental fragments it may be assumed that an oceanic region existed between them. After collision the former boundaries merge into a single continental region. The line (or lines) of joining together have been arbitrarily estimated on the maps and are shown by lines of crosses. The boundary between eastern Eurasia and North America has been drawn arbitrarily through the Bering Strait. A better choice might have been through the Verkhoyansk, Cherskiy or Chukchi mountain ranges.

The second source of uncertainty lies in the estimate of the mean paleomagnetic pole of the continental reassembly, which in turn depends on the accuracy of individual paleomagnetic pole estimates. The number N in the caption of each map refers to the number of separate paleomagnetic studies that have been used to make the map. A study has been accepted provided that it satisfies certain criteria (McElhinny, 1973), and provided too that it has an age

range lying within 10 Ma of the time for which the map is required. For example, the 40-Ma map draws on all reliable pole studies on the stable parts of the continents whose age ranges include some part of the interval 30–50 Ma. Poles that lie within deformed regions have been excluded. The age criterion for accepting poles has some disadvantages. In particular, it means that a pole with a poorly determined age range will appear in the pole list for a much larger number of maps than one with a precisely determined age range. It would be better to weight poles according to the precision of their determined ages, but we have not done this, principally because the appropriate statistics are not yet well developed. Nor are the poles weighted according to their own internal precision: imprecisely determined poles are weighted equally with precisely determined ones. We have refrained from weighting the data because, quite apart from the difficulty of designing an appropriate weighting scheme, it would introduce an element of regional bias. This is because paleomagnetic studies in high paleolatitudes yield, on average, less precise estimates of paleomagnetic poles as a consequence of the variation of magnetic inclination I with the latitude λ of a dipole field:

$$\tan I = 2 \tan \lambda.$$

An error dI in inclination gives rise to an error $d\lambda$ in the location of the pole on the paleomeridian:

$$d\lambda = \tfrac{1}{2}(1 + 3\sin^2\lambda)dI,$$

and this uncertainty increases with paleolatitude.

Alpha-95 in the caption of each map is a standard statistical measure for the spread of data on a sphere (Fisher, 1953). Essentially it is the radius in degrees of the circle of confidence around the calculated mean pole. The 95 per cent level means that there is a one in twenty chance that the true mean lies outside the circular limit. In this section, all the confidence circles have radii smaller than ten degrees; some are less than five degrees. This is a small dispersion for paleomagnetic data, though the estimated dispersion may be misleadingly small in some cases. This situation arises where most of the poles come from only one or two continental fragments; the effect of plate positioning errors is then less than when the data are more uniformly distributed among all the fragments. Also, each pole determination rather than each continental fragment has been weighted equally. Nevertheless, the scatter of the poles is remarkably small, and gives some idea of the uncertainties in the orientation of the reassembly.

Another source of error, which affects the construction both of the reassembly and of the map, is the interrelationship between the fossil, magnetic reversal and isotopic timescales. Much of the raw data used has been collected from the literature of the past fifteen years, during which a variety of conflicting timescales has been proposed. For the magnetic reversal timescale we have used that of Heirtzler *et al.* (1968), updated according to La Brecque, Kent & Cande (1977). The ages in Ma of the paleomagnetic poles have been taken as their isotopic age (where known), or calculated by converting the stratigraphic range in the pole compilation, using a combination of the timescales of Harland, Smith & Wilcock (1964) and Van Eysinga (1975). The latter timescale places the Paleozoic–Mesozoic boundary at 232 Ma instead of 225 Ma. This effectively lengthens the age in Ma assigned to the Triassic period. The geological series names in the map captions have been taken from the Van Eysinga timetable.

The total effect of all the above sources of error is not known and is difficult to estimate.

Map 1

Present day

⊞ Cenozoic orogenic belts

⊡ Mesozoic orogenic belts

▨ younger Paleozoic orogenic belts

■ older Paleozoic orogenic belts

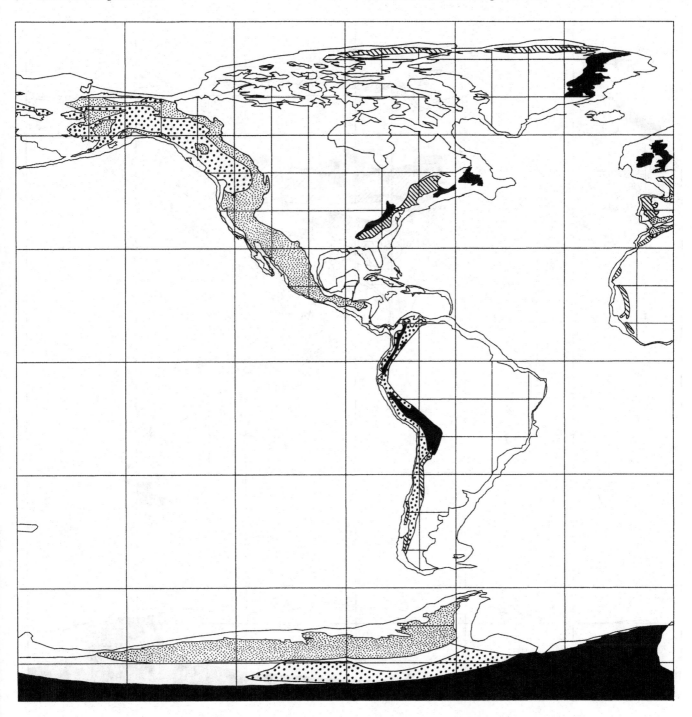

Present day
- ▦ Cenozoic orogenic belts
- ⊡ Mesozoic orogenic belts

- ▨ younger Paleozoic orogenic belts
- ■ older Paleozoic orogenic belts

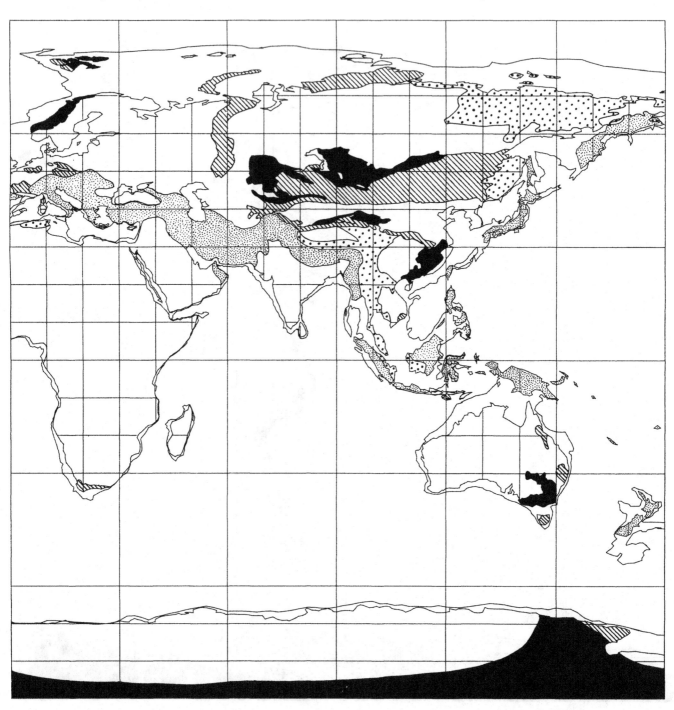

Map 3
Present day

North polar Lambert equal-area

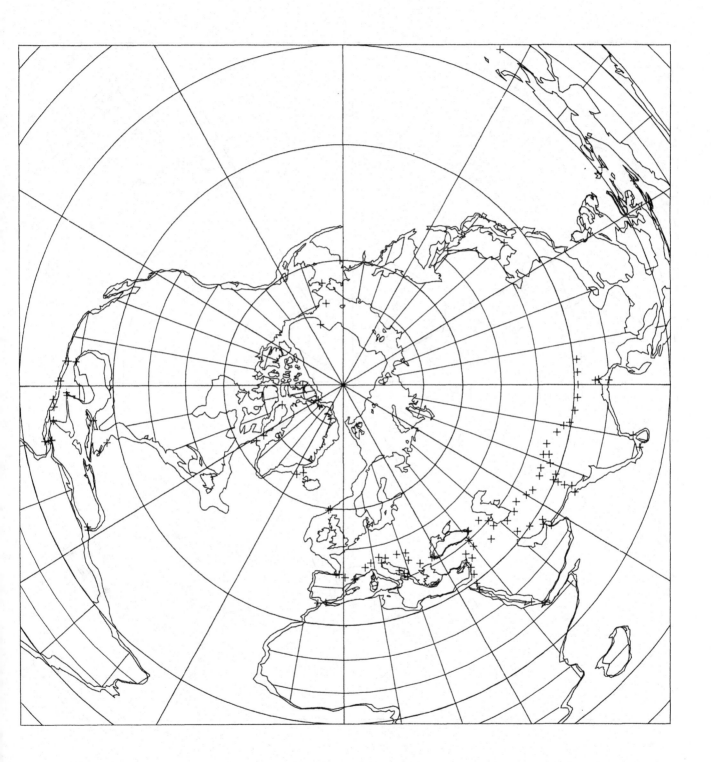

Map 4
Present day

South polar Lambert equal-area

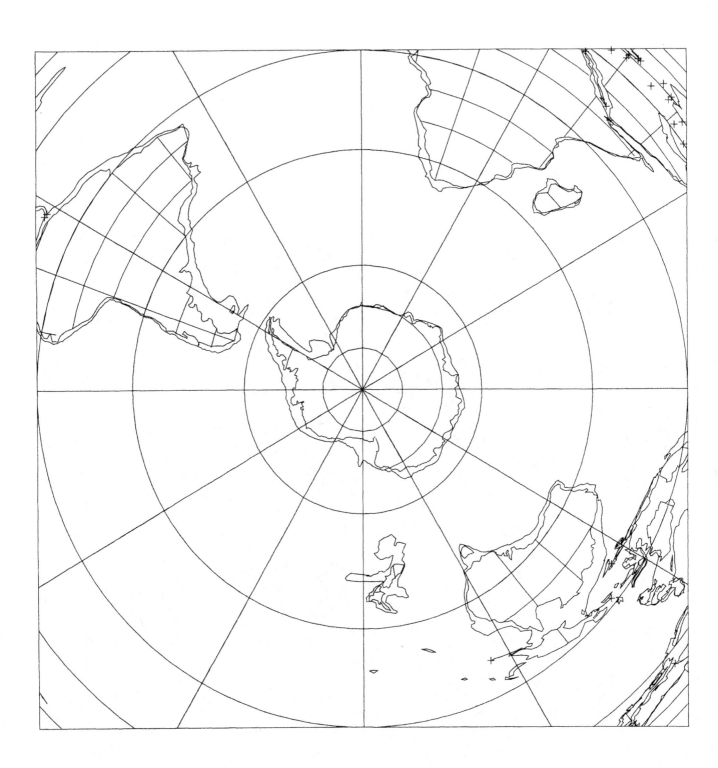

Map 5
10 million years
late Miocene (Cenozoic)

Cylindrical equidistant
$N = 172$ alpha-95 = 1.8

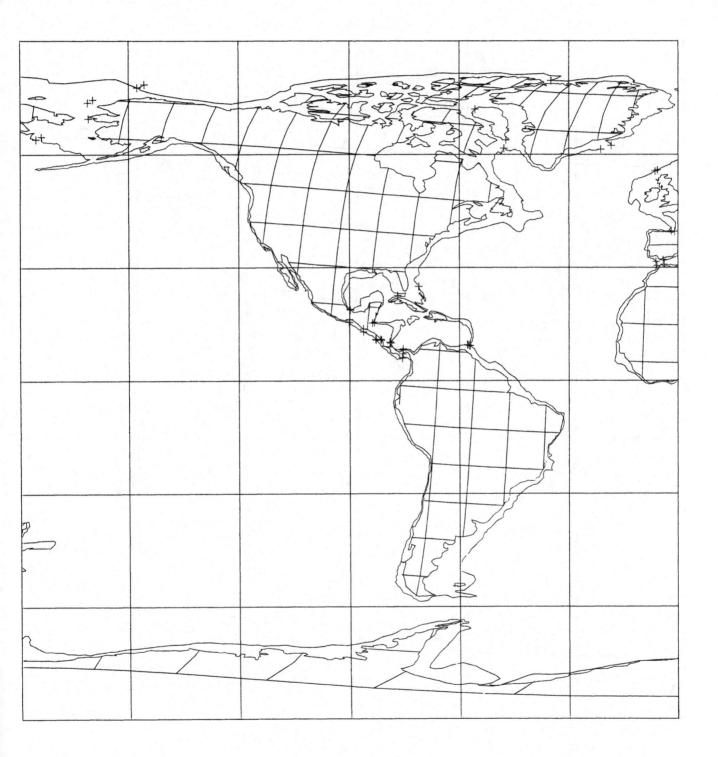

Map 6
10 million years
late Miocene (Cenozoic)

Cylindrical equidistant
$N = 172$ alpha-95 = 1.8

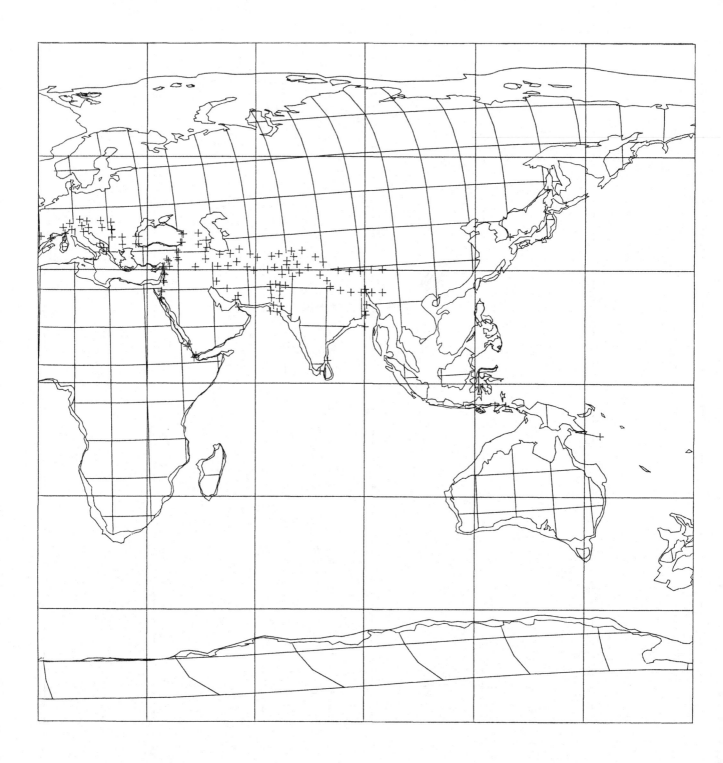

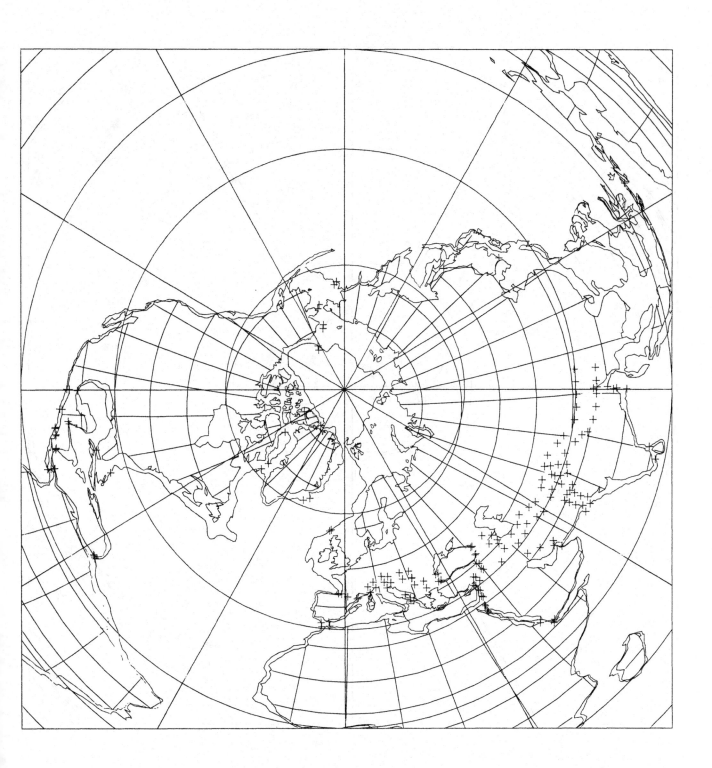

13

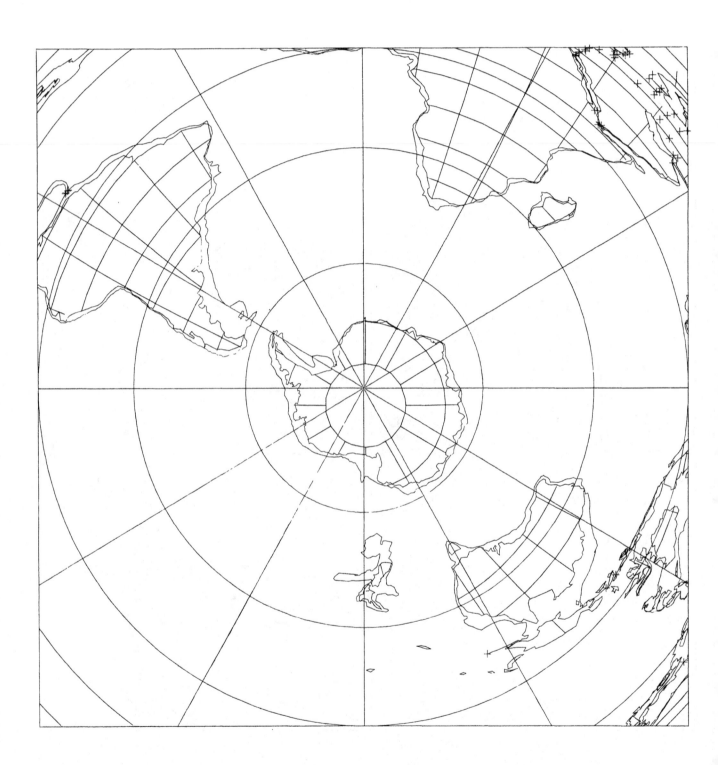

Map 9
20 million years
early Miocene (Cenozoic)

Cylindrical equidistant
$N = 89$ alpha-95 = 2.5

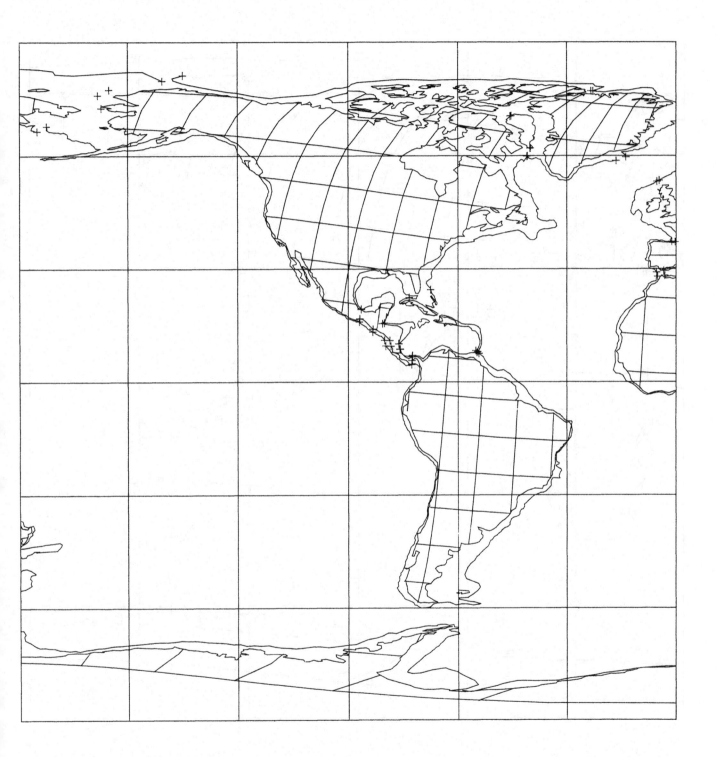

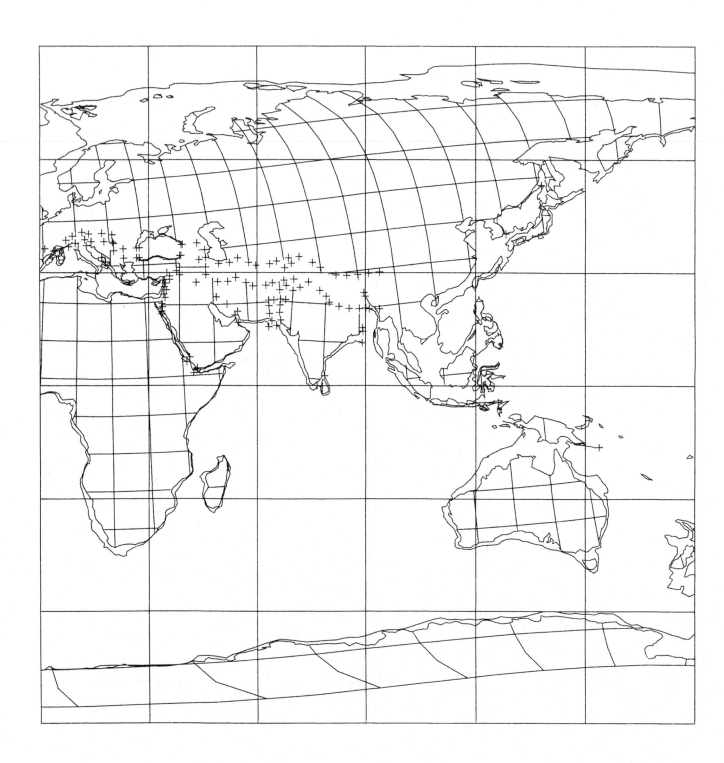

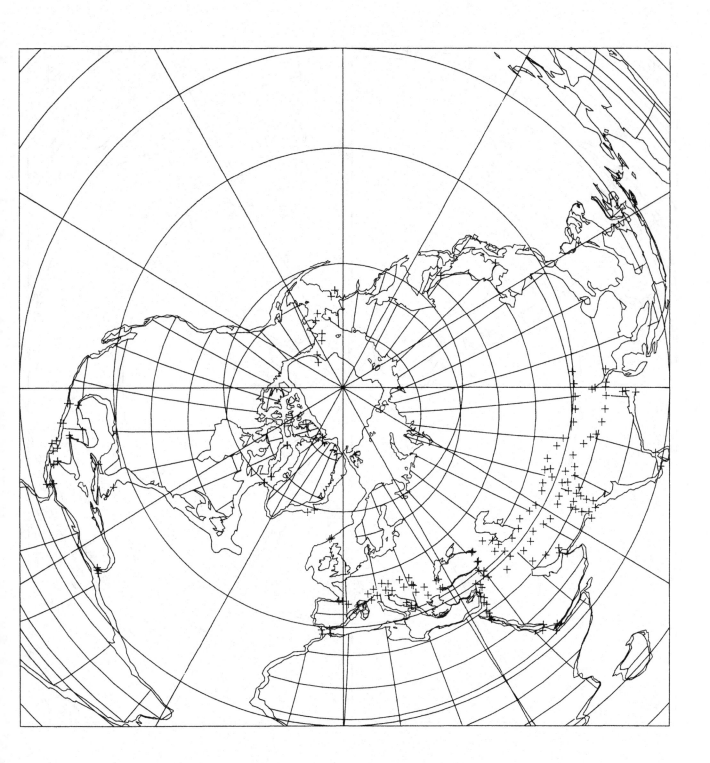

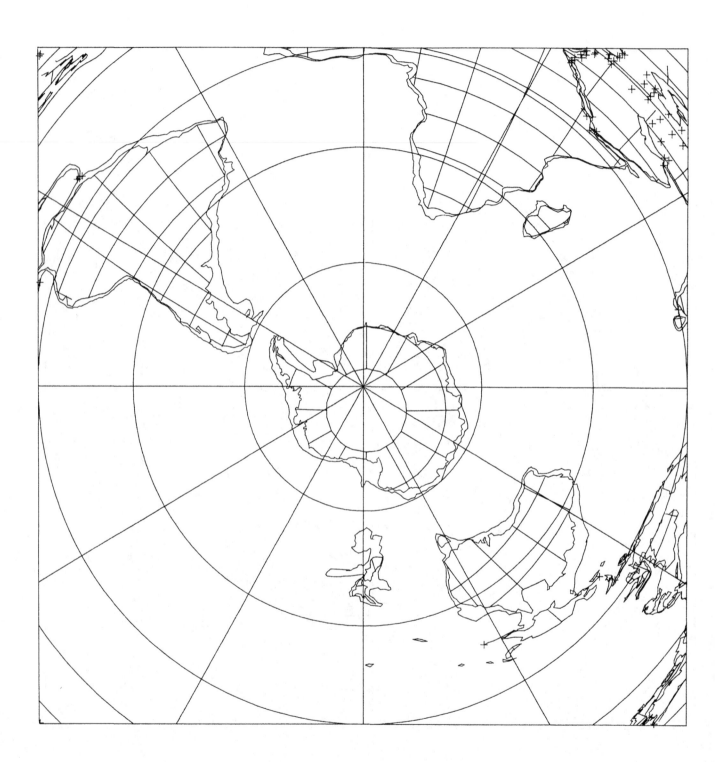

Map 13
40 million years
late Eocene (Cenozoic)

Cylindrical equidistant
$N = 60$ alpha-95 = 4.4

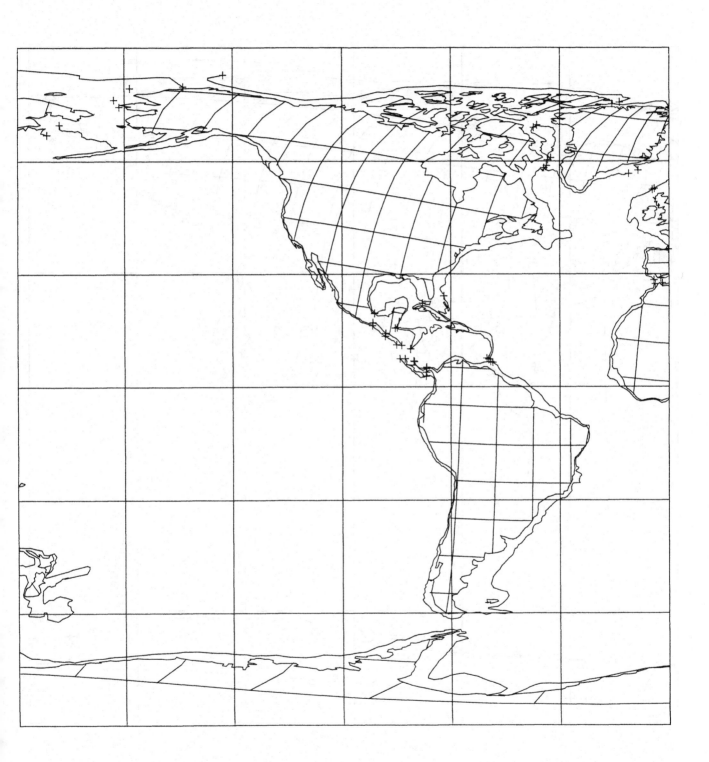

Map 14
40 million years
late Eocene (Cenozoic)

Cylindrical equidistant
$N = 60$ alpha-95 = 4.4

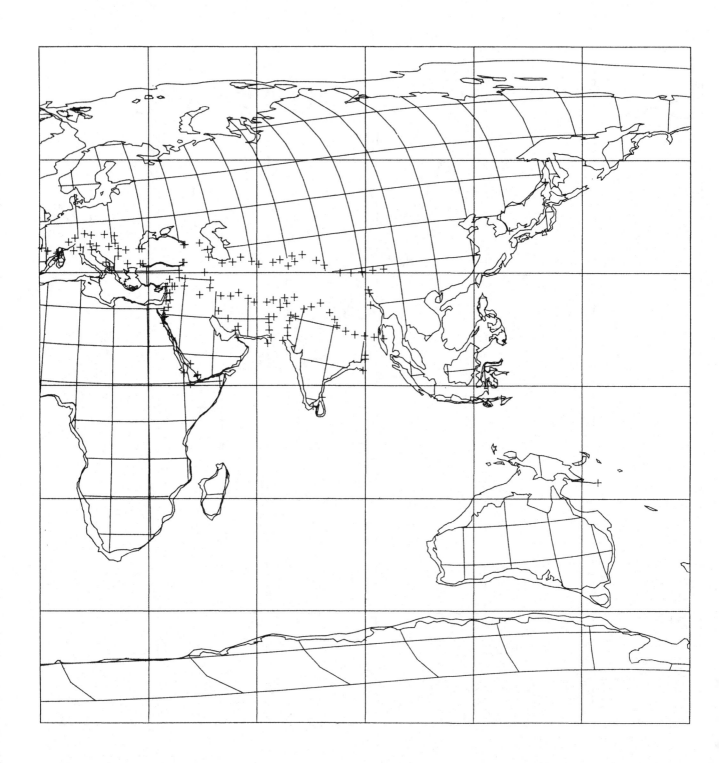

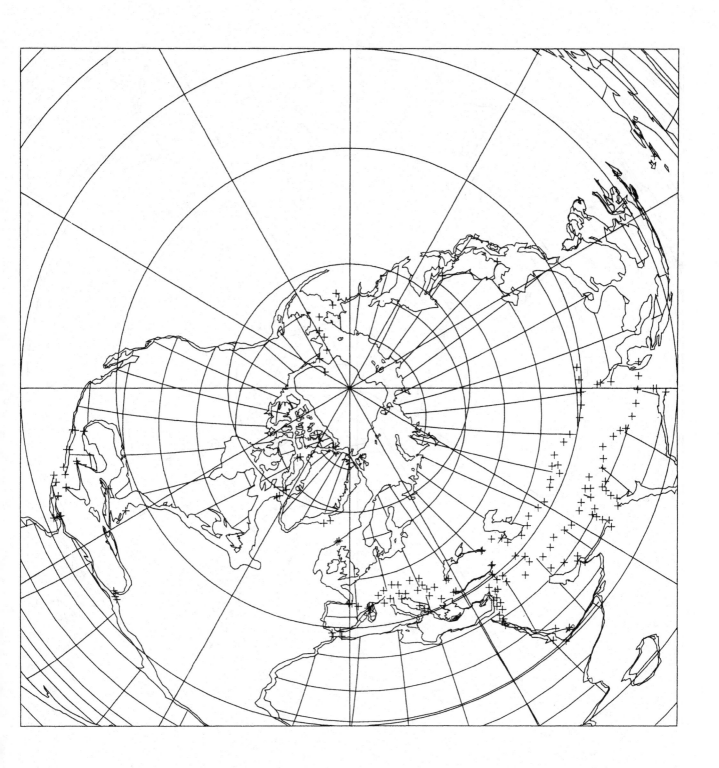

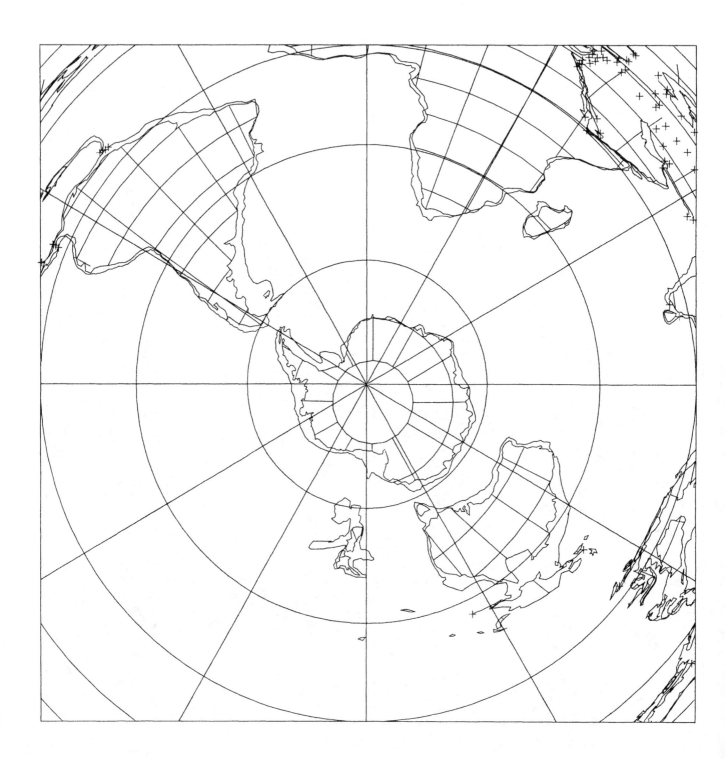

Map 17
60 million years
Paleocene (Cenozoic)

Cylindrical equidistant
$N = 94$ alpha-95 = 3.6

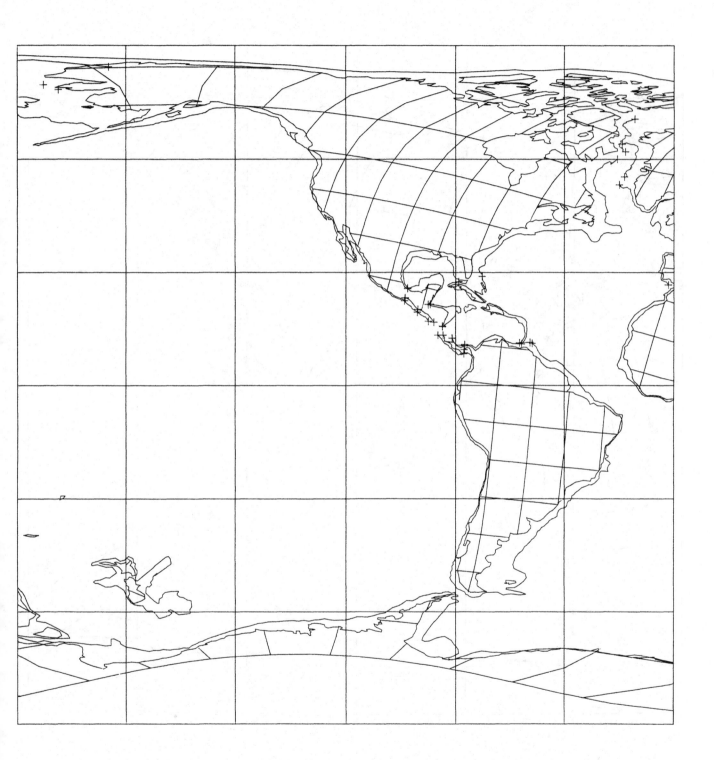

Map 18
60 million years
Paleocene (Cenozoic)

Cylindrical equidistant
$N = 94$ alpha-95 = 3.6

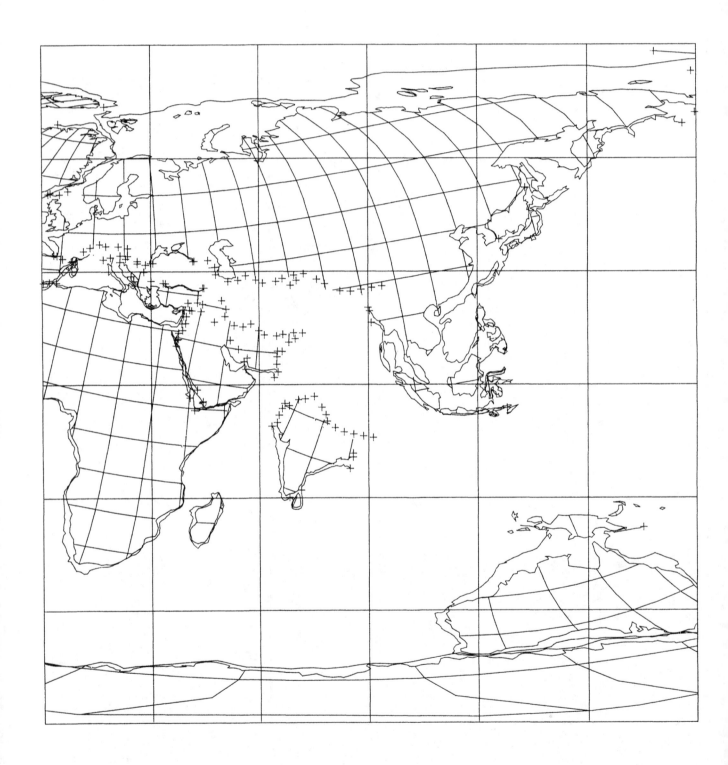

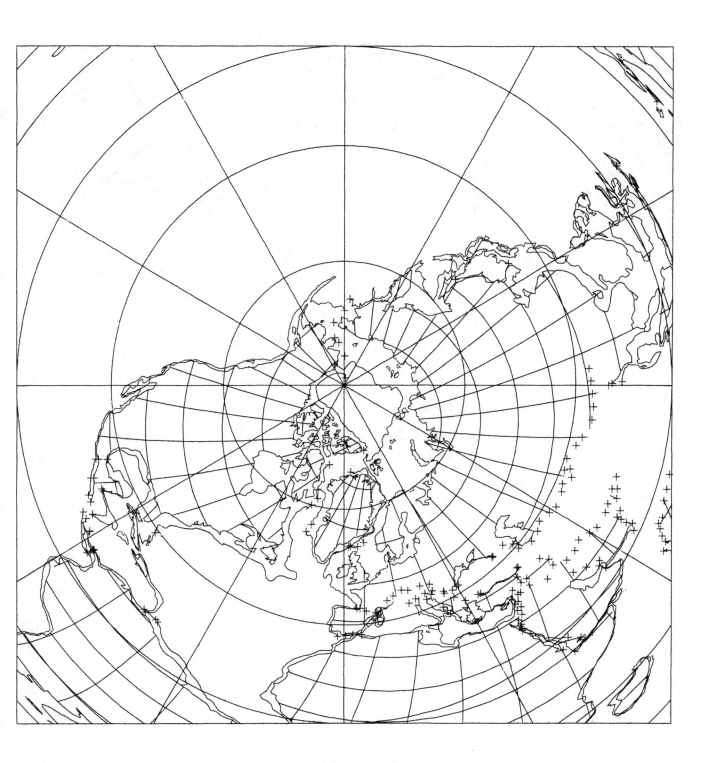

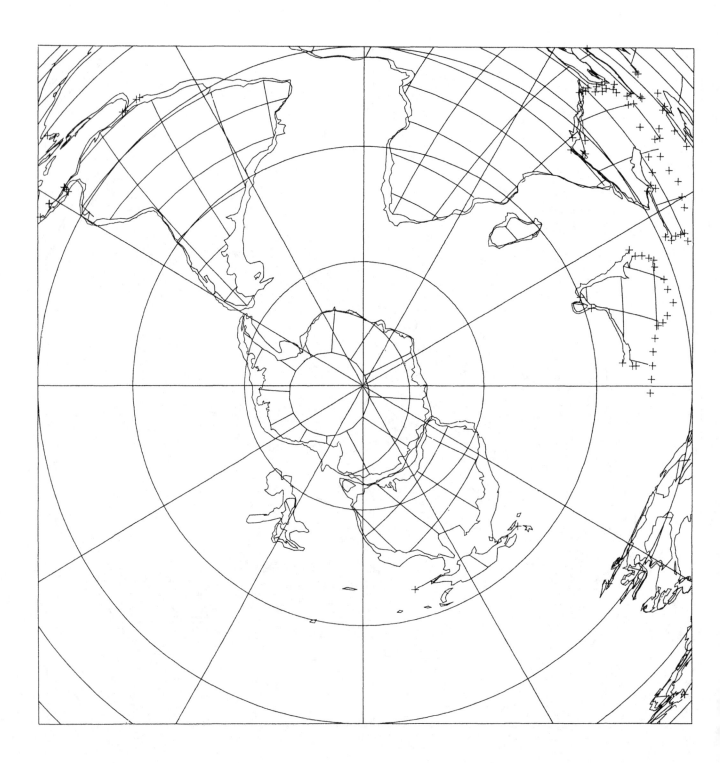

Map 21
80 million years
Santonian (late Cretaceous)

Cylindrical equidistant
$N = 63$ alpha-95 = 4.8

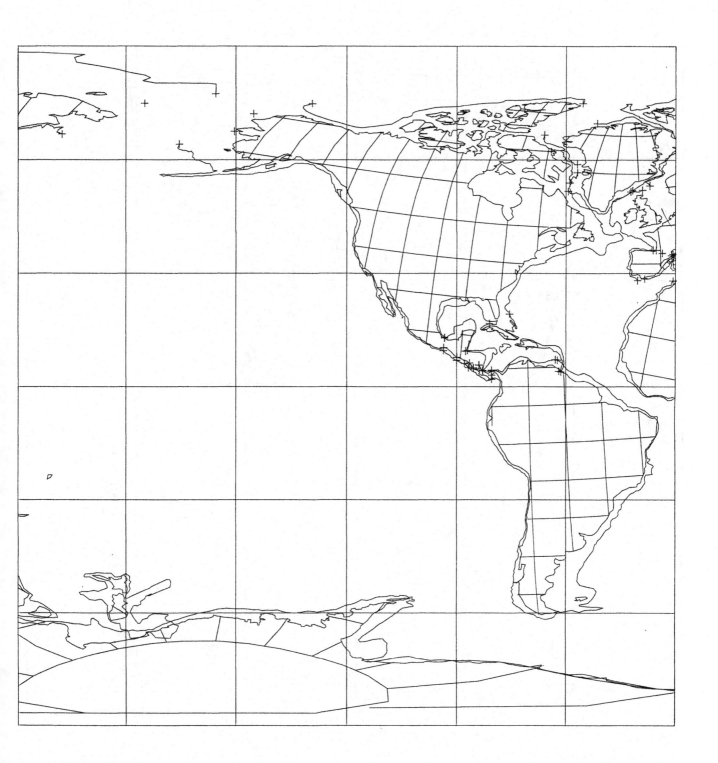

27

Map 22
80 million years
Santonian (late Cretaceous)

Cylindrical equidistant
$N = 63$ alpha-95 = 4.8

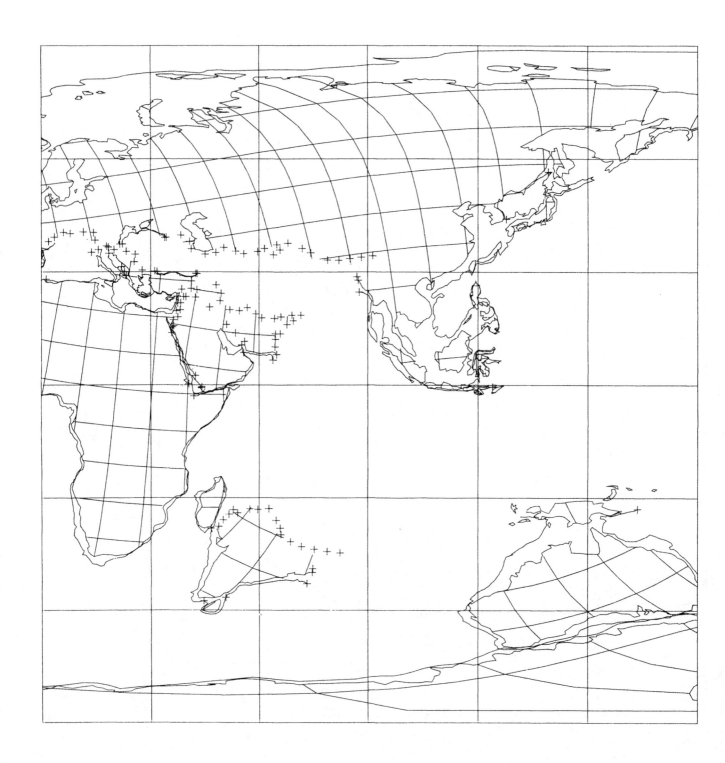

Map 23
80 million years
Santonian (late Cretaceous)

North polar Lambert equal-area
$N = 63$ alpha-95 = 4.8

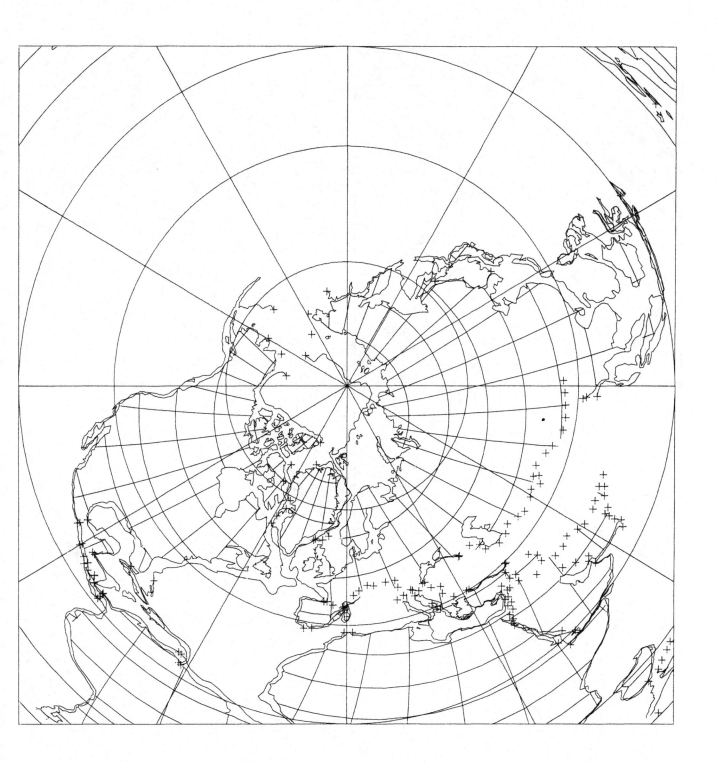

Map 24
80 million years
Santonian (late Cretaceous)

South polar Lambert equal-area
$N = 63$ alpha-95 = 4.8

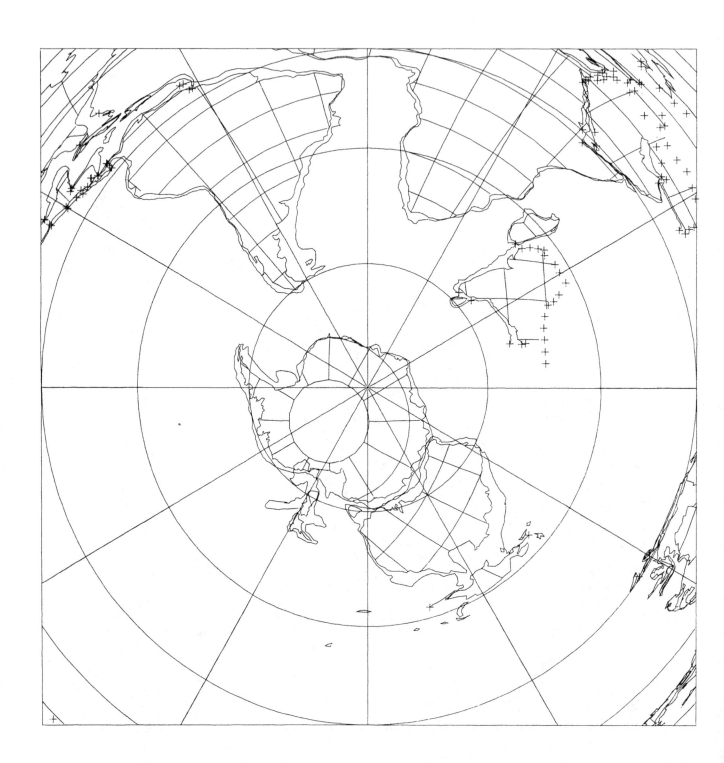

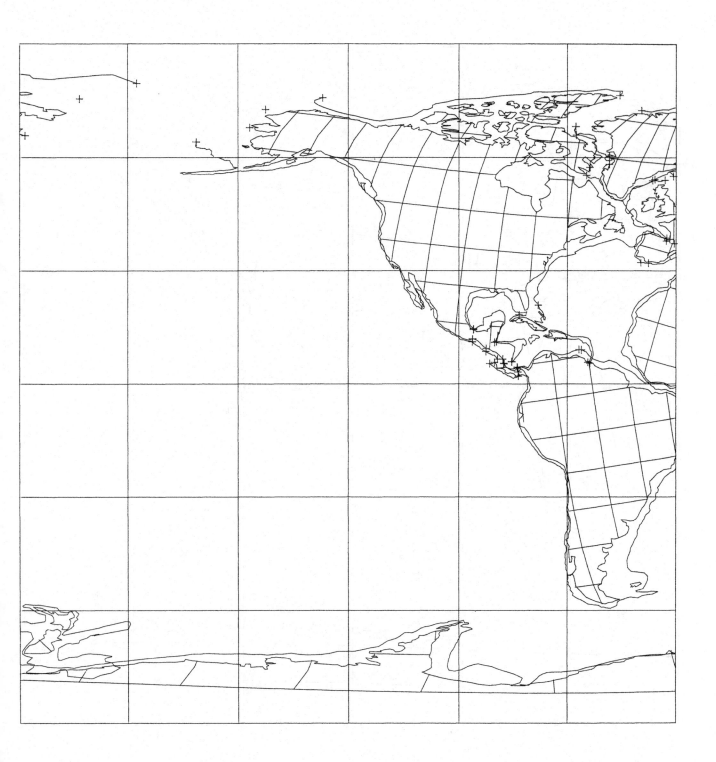

Map 26
100 million years
earliest Cenomanian (mid Cretaceous)

Cylindrical equidistant
$N = 70$ alpha-95 = 4.5

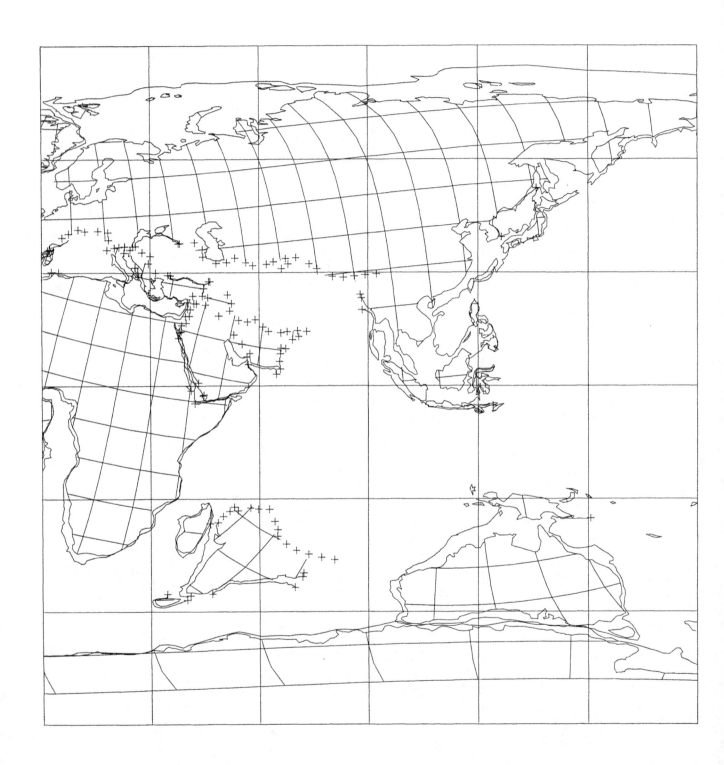

Map 27
100 million years
earliest Cenomanian (mid Cretaceous)

North polar Lambert equal-area
$N = 70$ alpha-95 = 4.5

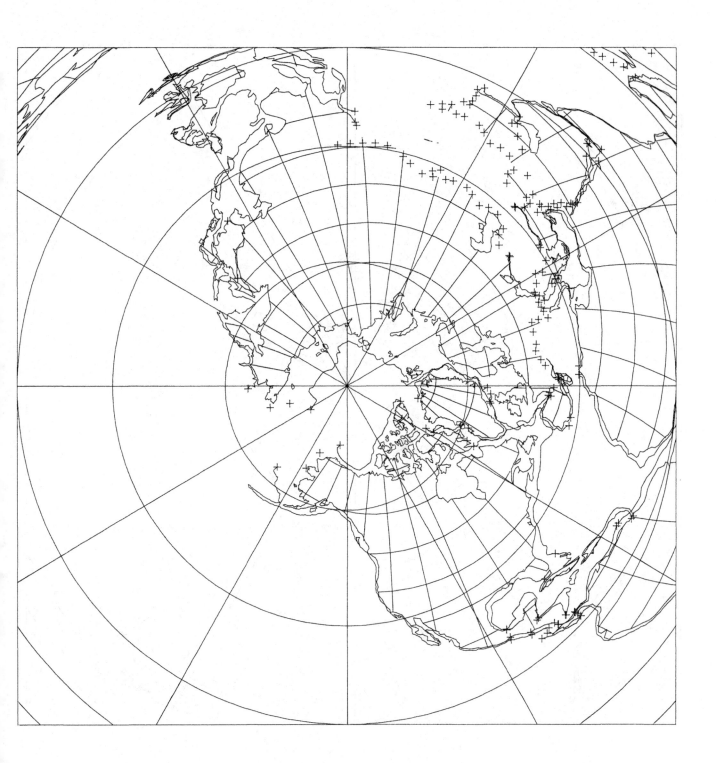

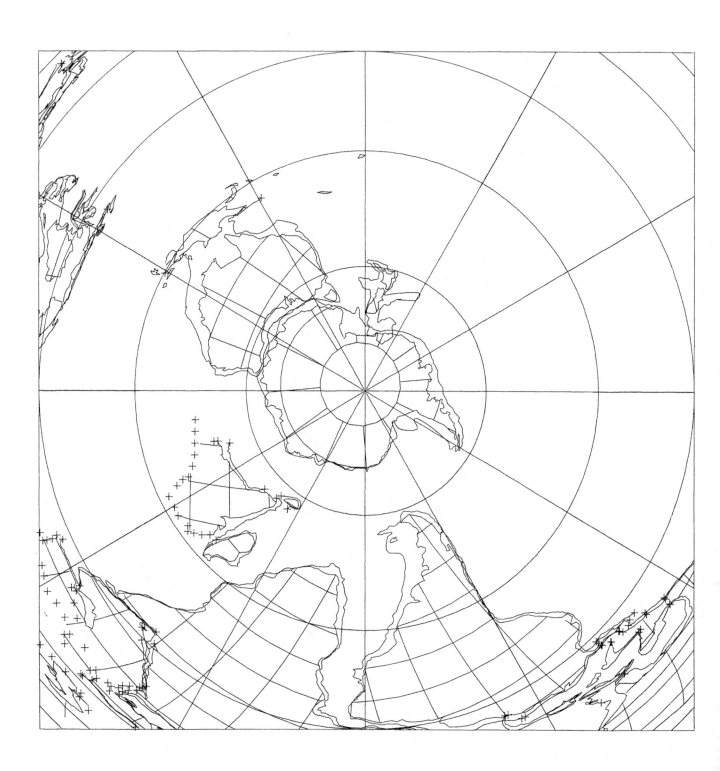

Map 29
120 million years
Hauterivian (early Cretaceous)

Cylindrical equidistant
$N = 44$ alpha-95 = 5.6

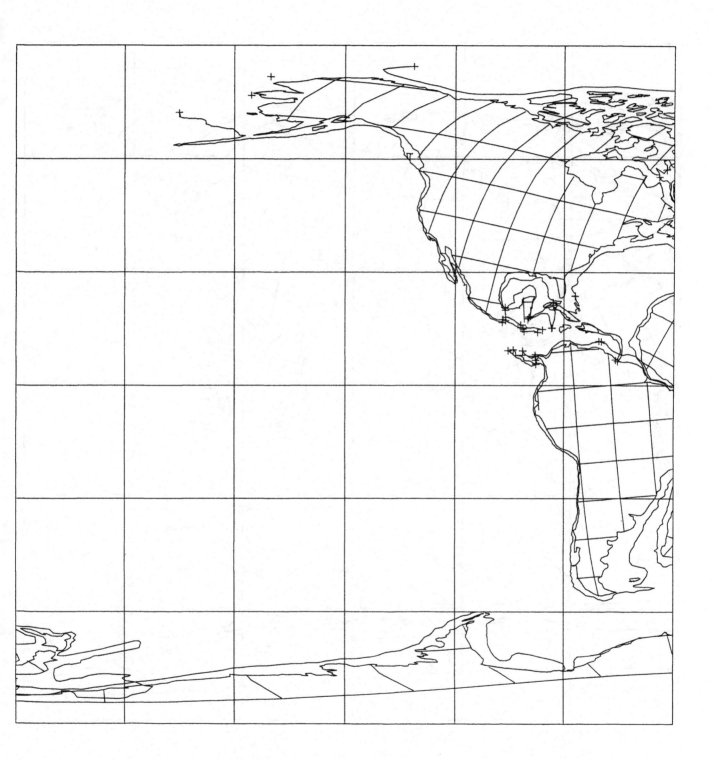

Map 30
120 million years
Hauterivian (early Cretaceous)

Cylindrical equidistant
$N = 44$ alpha-95 = 5.6

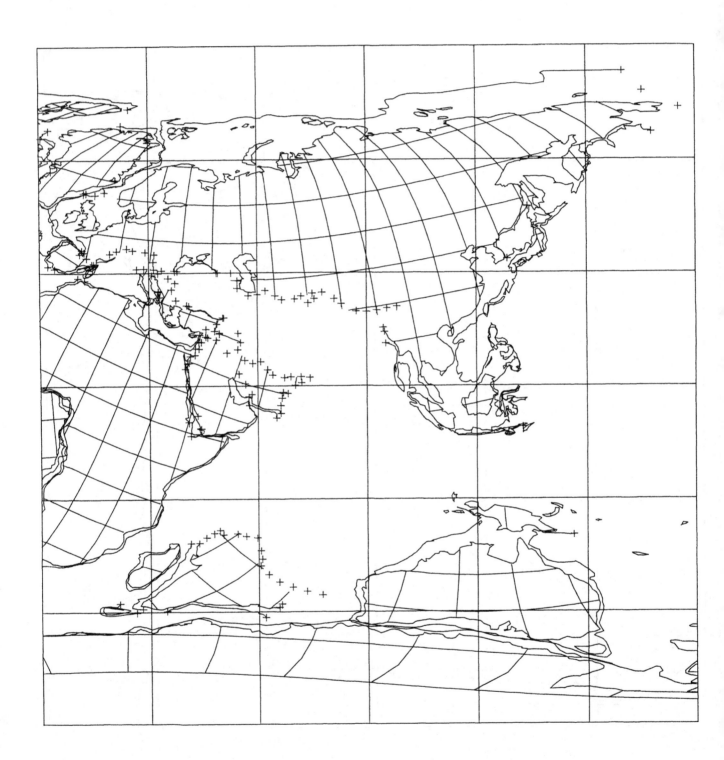

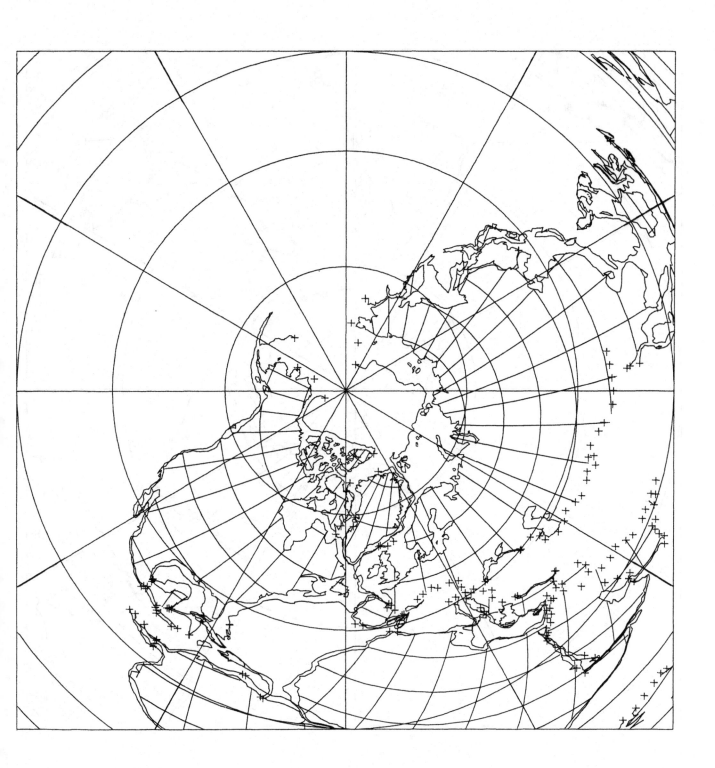

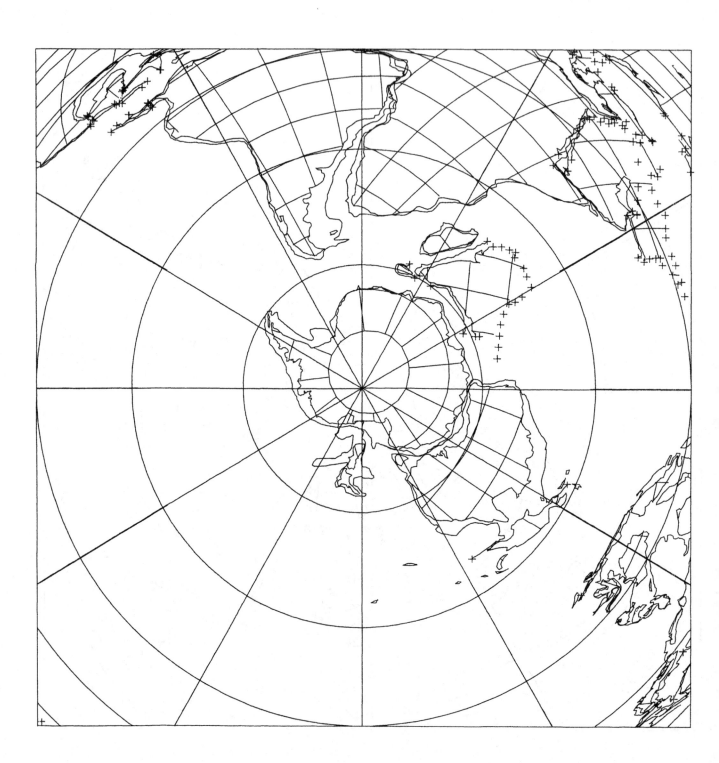

Map 33
140 million years
'Tithonian' (late Jurassic)

Cylindrical equidistant
$N = 58$ alpha-95 = 5.6

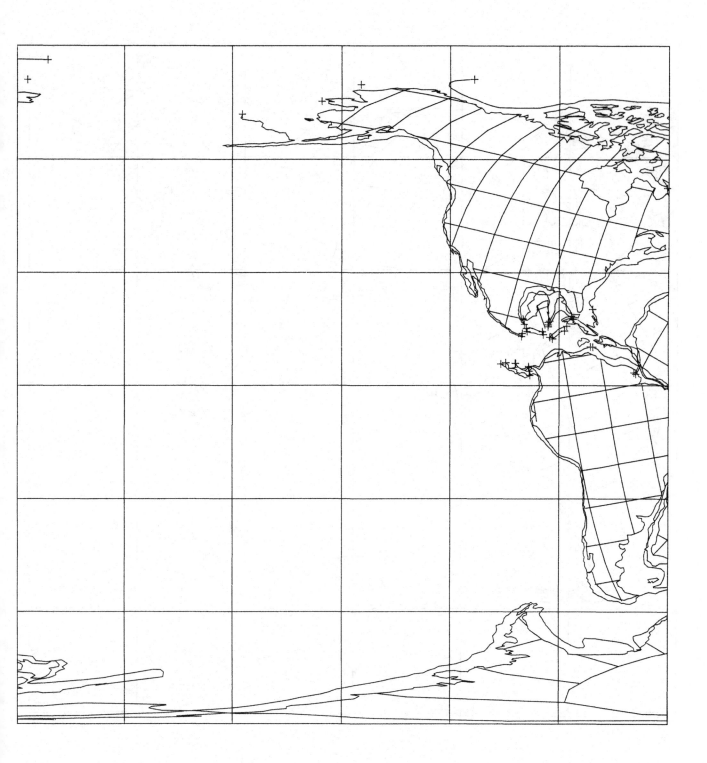

Map 34
140 million years
'Tithonian' (late Jurassic)

Cylindrical equidistant
$N = 58$ alpha-95 = 5.6

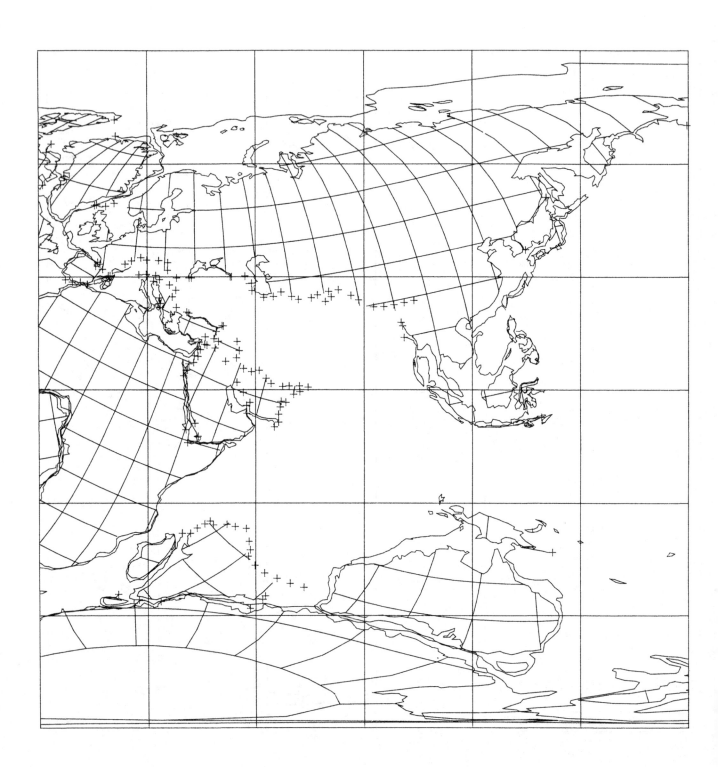

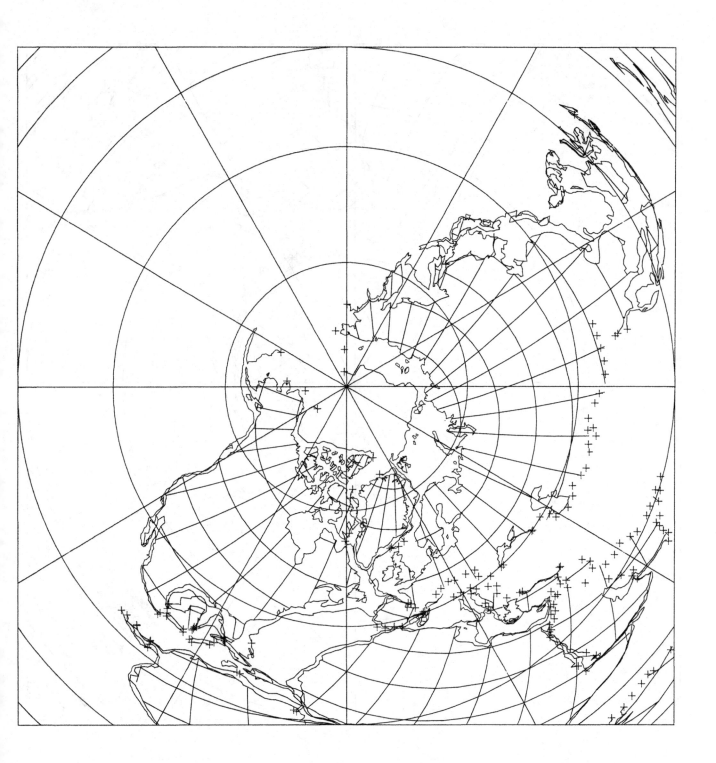

41

Map 36
140 million years
'Tithonian' (late Jurassic)

South polar Lambert equal-area
$N = 58$ alpha-95 = 5.6

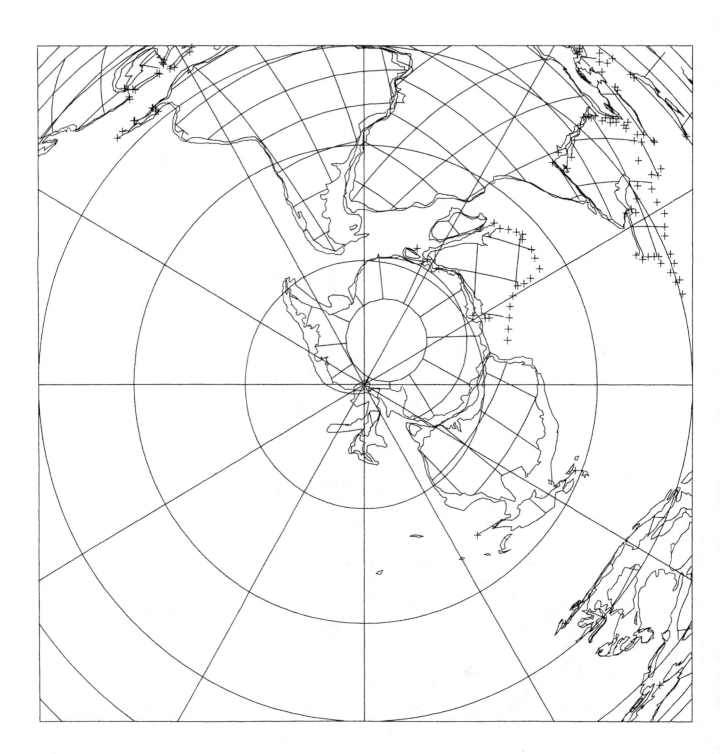

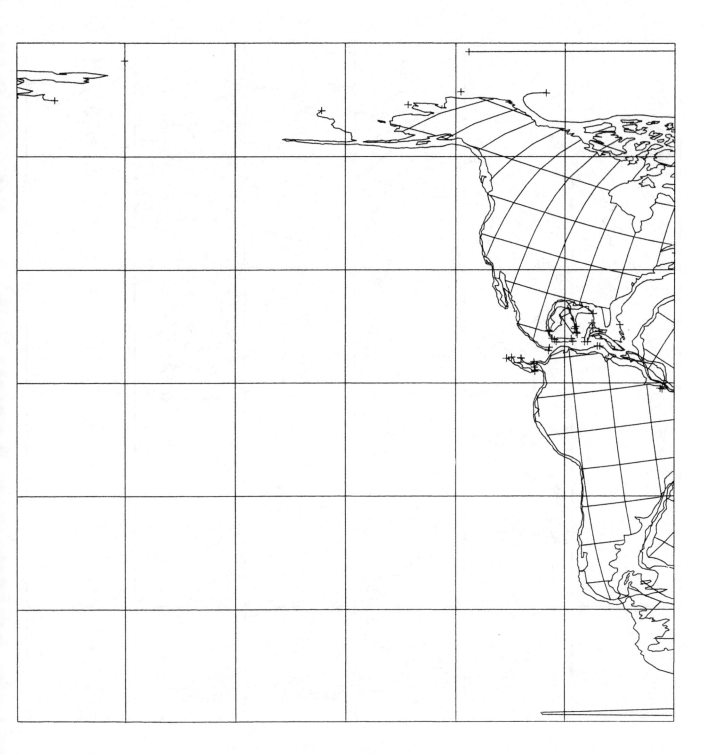

Map 38
160 million years
Callovian (mid Jurassic)

Cylindrical equidistant
$N = 41$ alpha-95 = 6.1

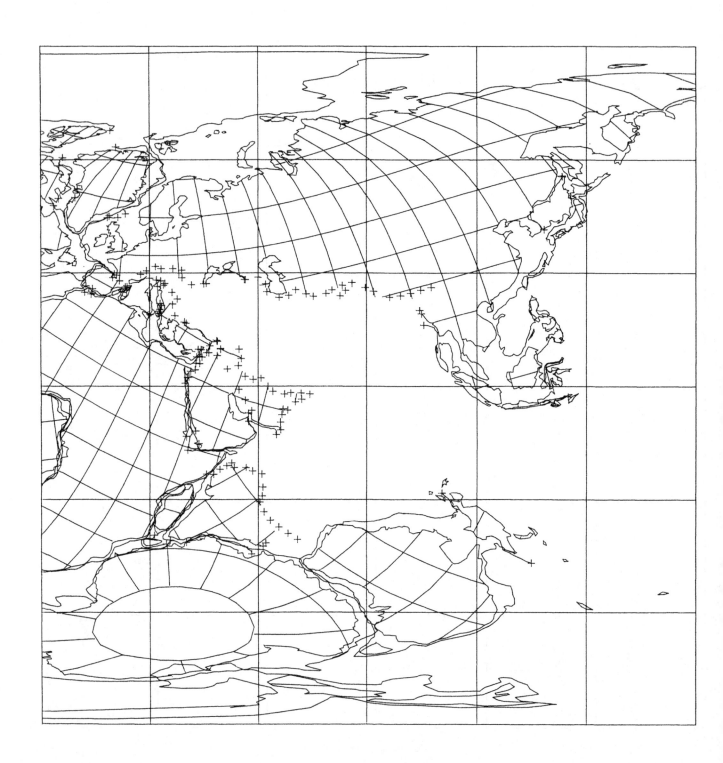

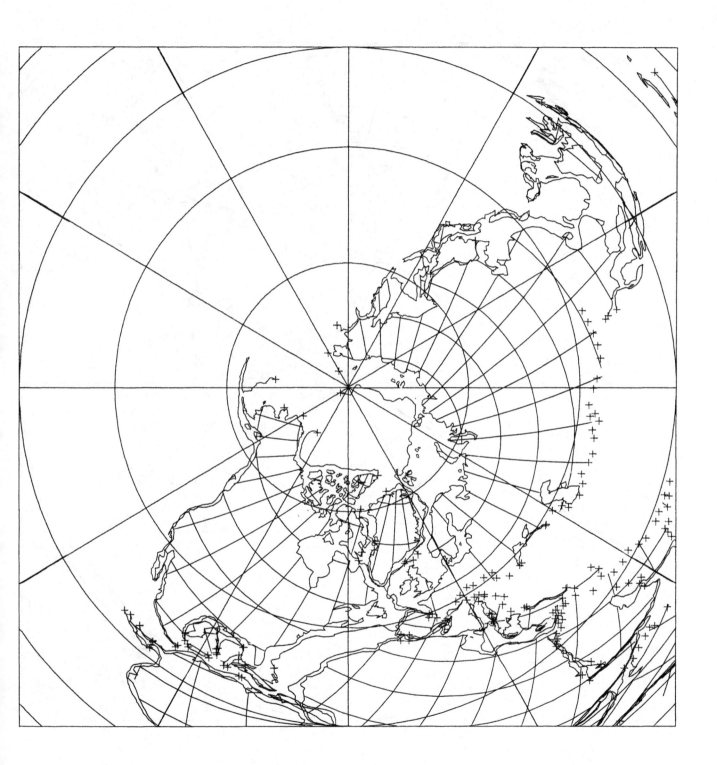

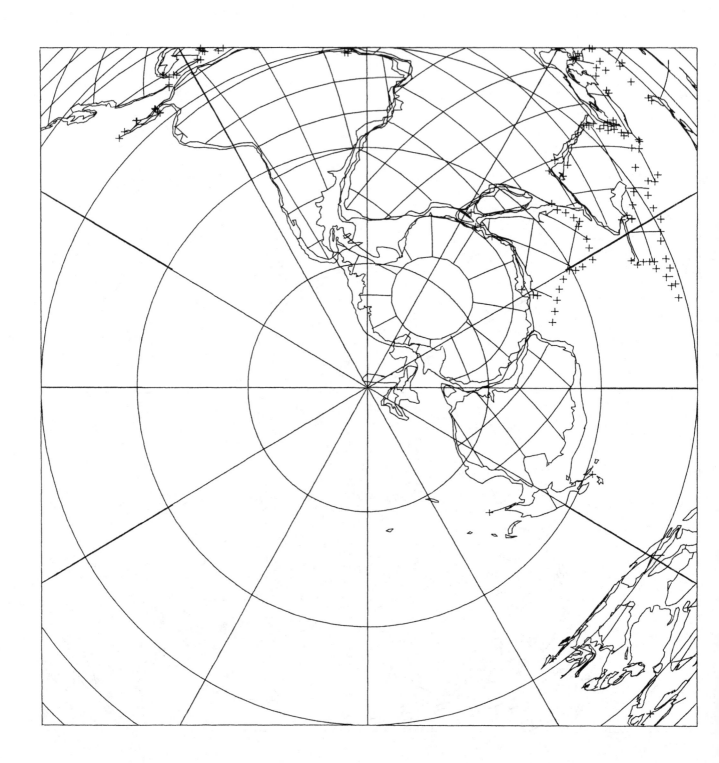

46

Map 41
180 million years
Pliensbachian (early Jurassic)

Cylindrical equidistant
$N = 63$ alpha-95 = 4.0

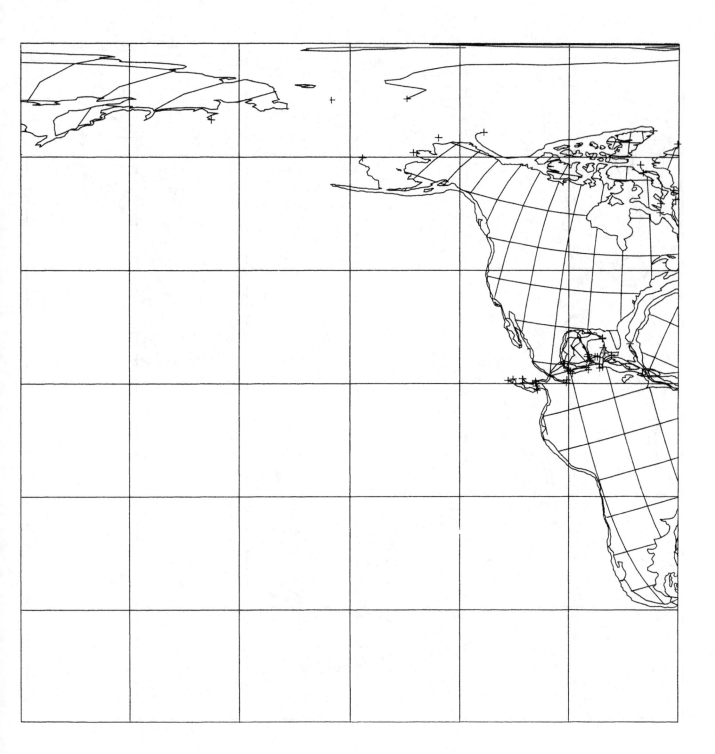

Map 42
180 million years
Pliensbachian (early Jurassic)

Cylindrical equidistant
$N = 63$ alpha-95 = 4.0

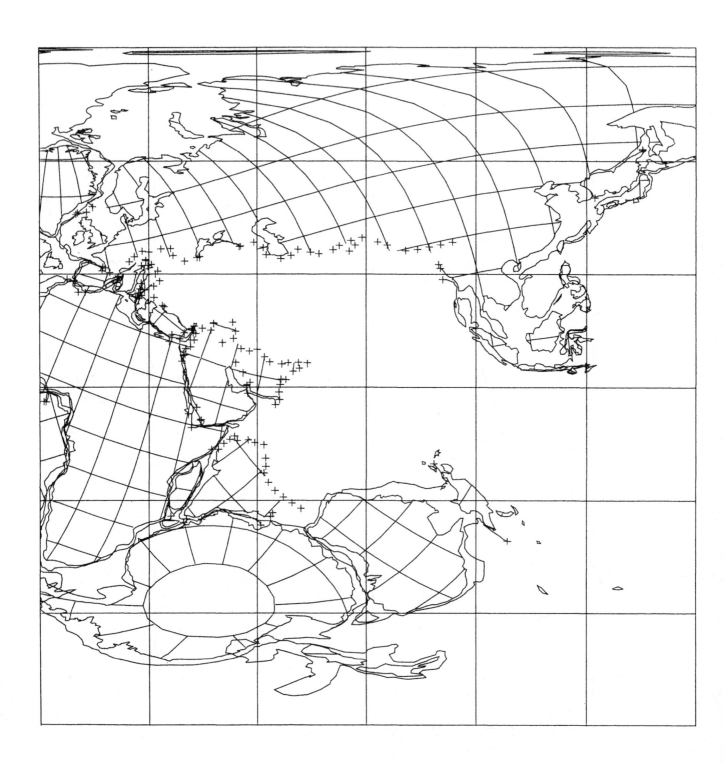

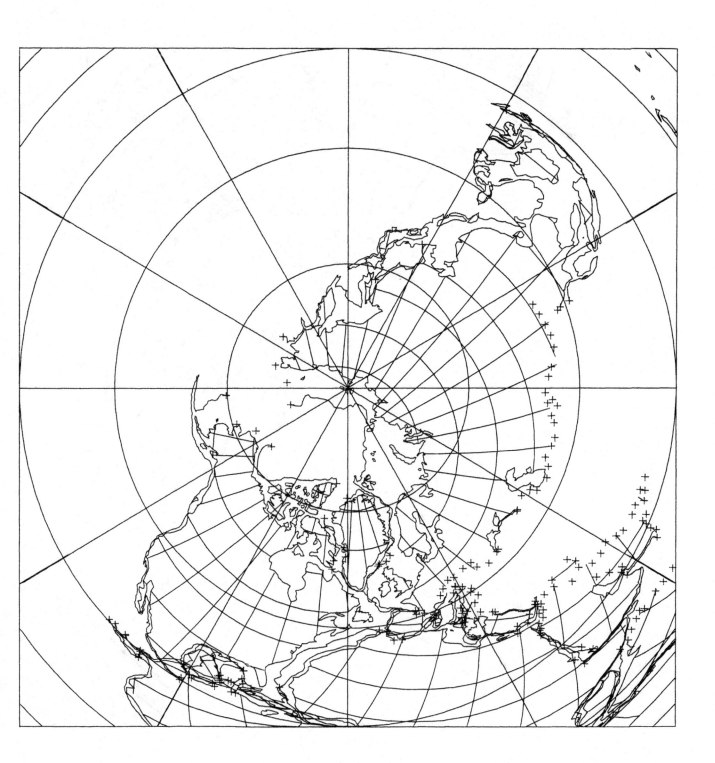

Map 44
180 million years
Pliensbachian (early Jurassic)

South polar Lambert equal-area
$N = 63$ alpha-95 = 4.0

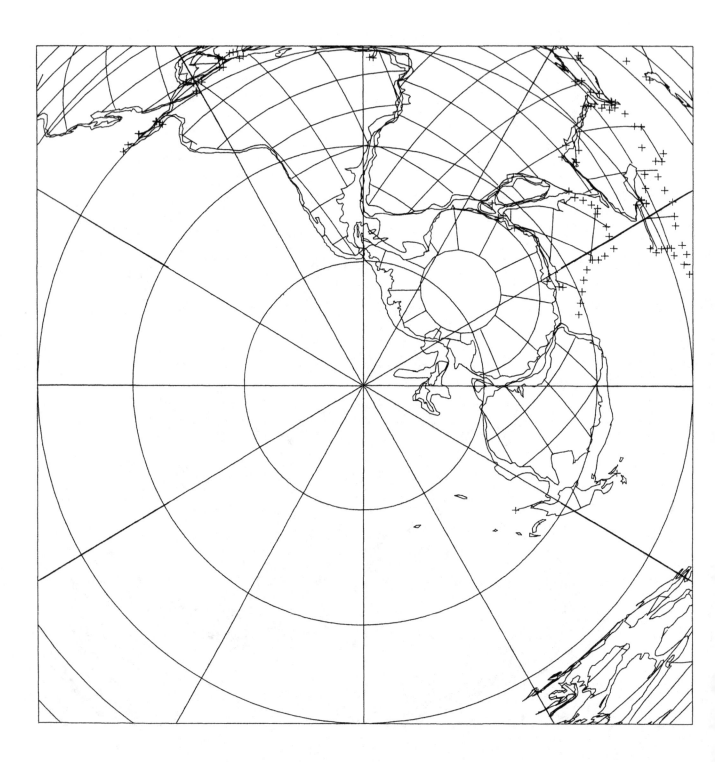

Map 45
200 million years
approx. Rhaetian (latest Triassic)

Cylindrical equidistant
$N = 81$ alpha-95 = 4.3

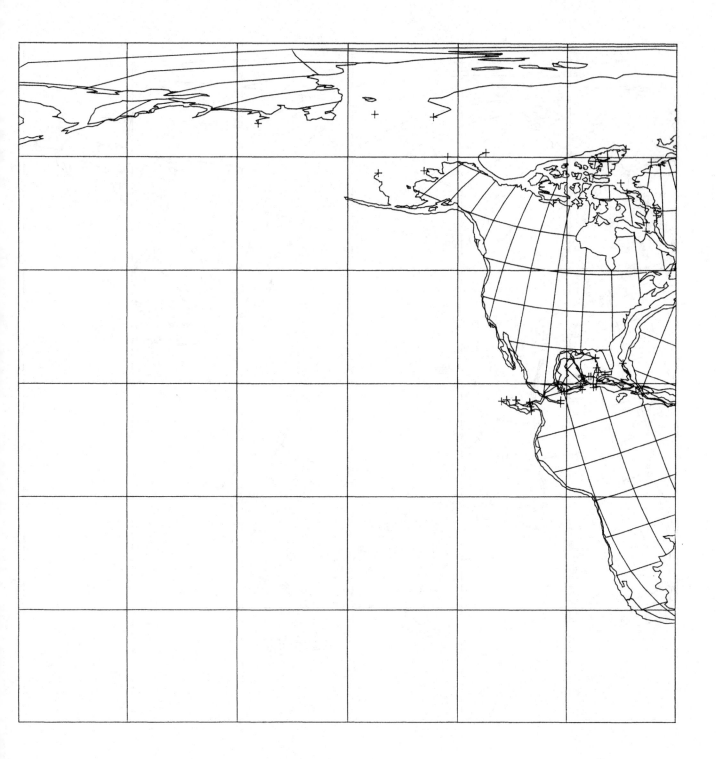

Map 46
200 million years
approx. Rhaetian (latest Triassic)

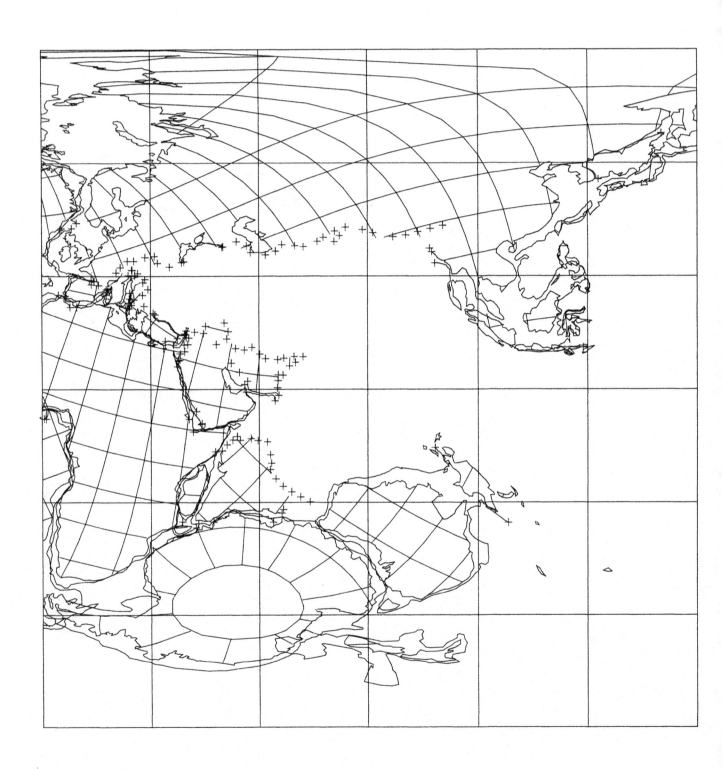

Map 47
200 million years
approx. Rhaetian (latest Triassic)

North polar Lambert equal-area
$N = 81$ alpha-95 = 4.3

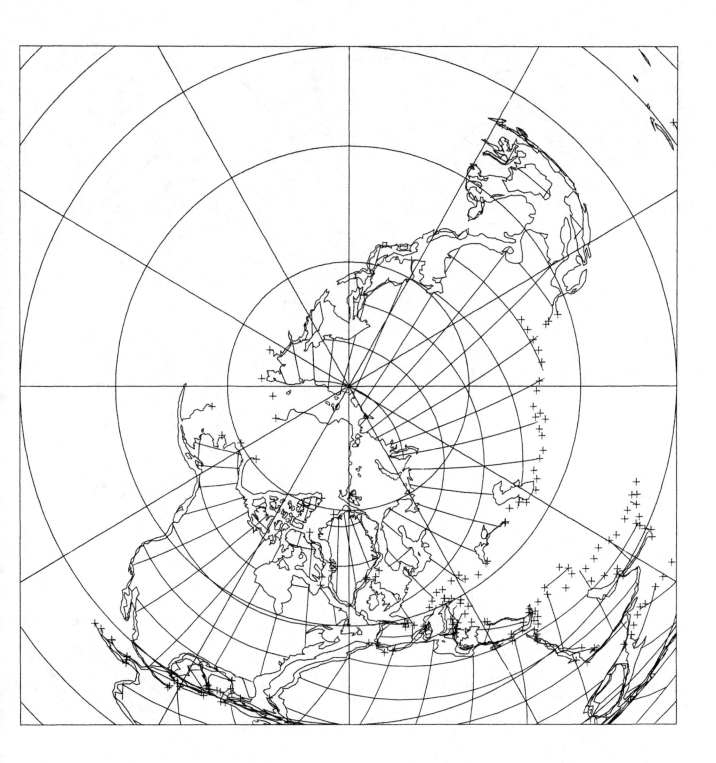

Map 48
200 million years
approx. Rhaetian (latest Triassic)

South polar Lambert equal-area
$N = 81$ alpha-95 = 4.3

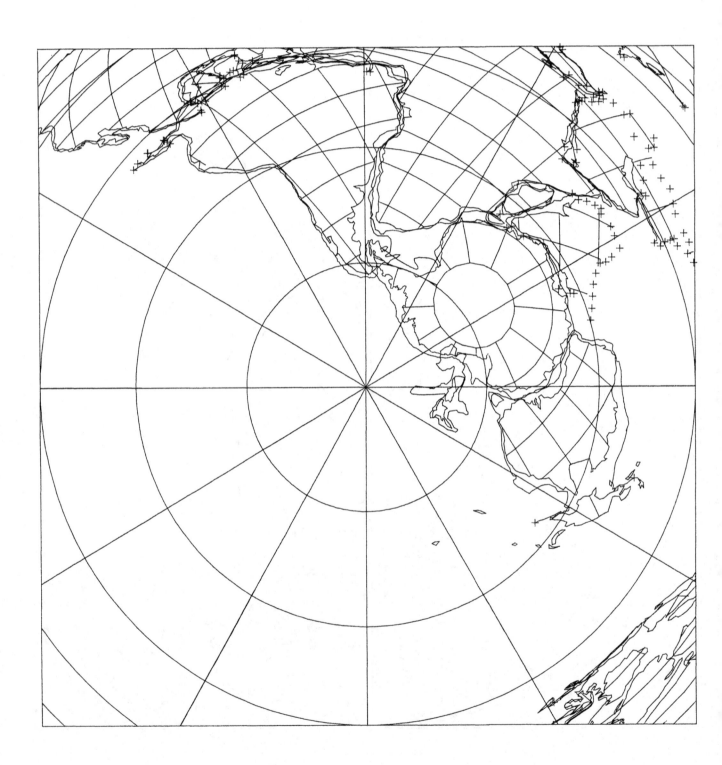

Map 49
220 million years
approx. Anisian (late Triassic)

Cylindrical equidistant
$N = 108$ alpha-95 = 3.6

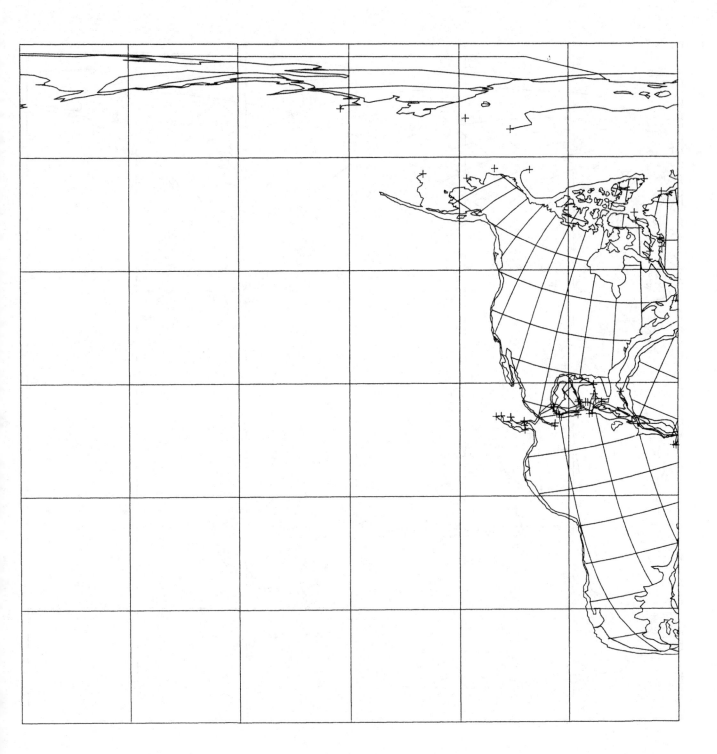

Map 50
220 million years
approx. Anisian (late Triassic)

Cylindrical equidistant
$N = 108$ alpha-95 = 3.6

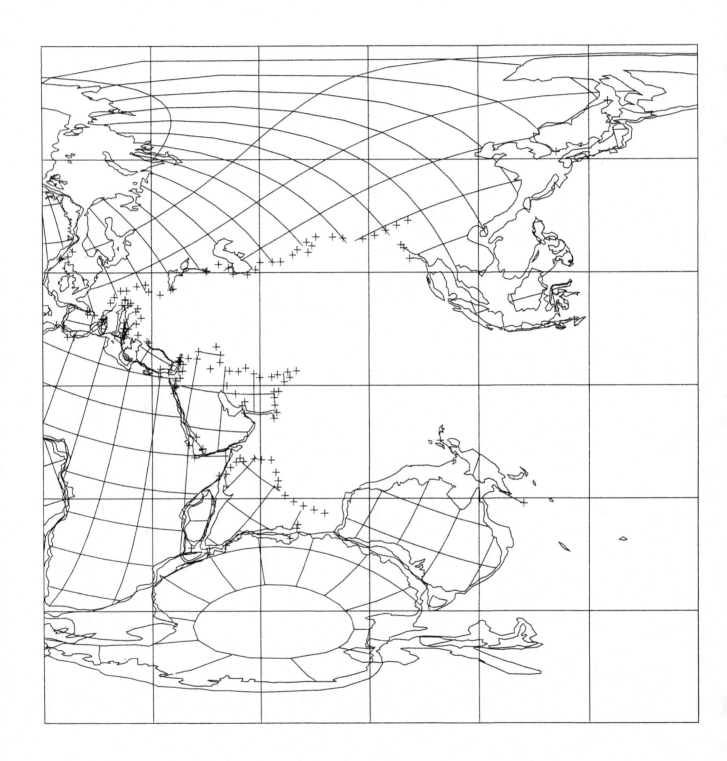

Map 51
220 million years
approx. Anisian (late Triassic)

North polar Lambert equal-area
$N = 108$ alpha-95 = 3.6

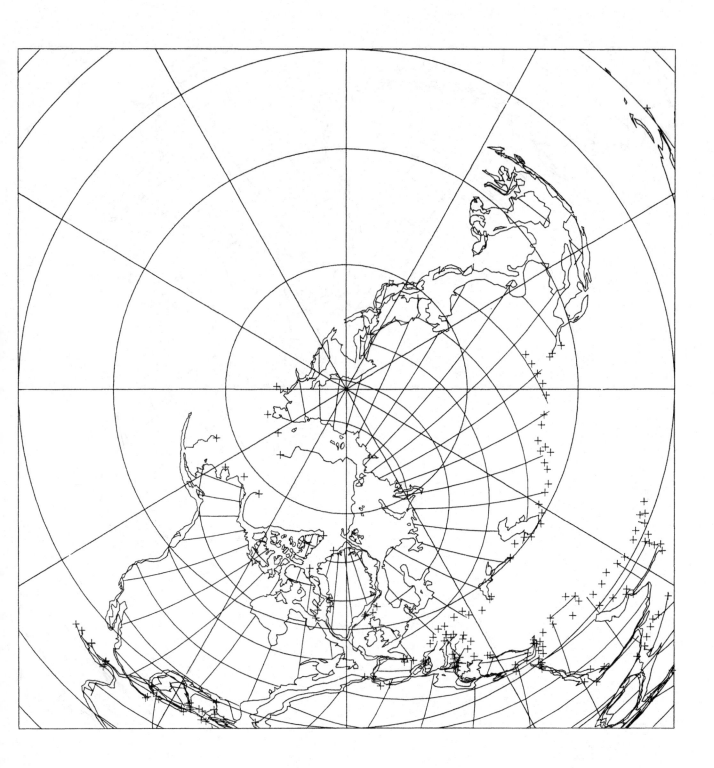

Map 52
220 million years
approx. Anisian (late Triassic)

South polar Lambert equal-area
$N = 108$ alpha-95 = 3.6

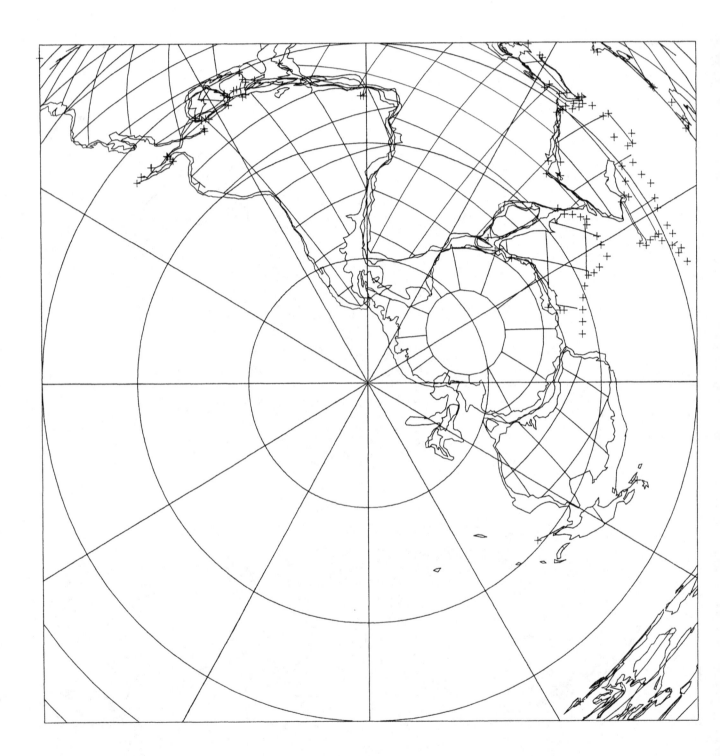

Section 2: Paleozoic composite maps

Introduction
This section contains thirty-six maps covering the time period 240 Ma to 560 Ma at 40-Ma intervals.

Method of making the composites
The main difference between the method of making these composites and the paleocontinental maps of section 1 is that each of the Paleozoic continents is projected separately as a fragment onto a global geographic frame. The frames are then superimposed so that no continental fragment overlaps any other. This method produces a composite of the continental fragments separated by arbitrary longitudes.

The main three continents which existed at the end of the Paleozoic were Gondwanaland, Laurasia and eastern Eurasia. It seems likely that Gondwanaland comprised Africa, South America, Australia, Antarctica, Arabia, India and Madagascar. Parts of present-day eastern Eurasia, such as Tibet, may also have been attached to Gondwanaland, but this possibility is not shown on any of the maps. We have also grouped Florida and the Caribbean region with Gondwanaland, the former based on paleontologic evidence, the latter because of geometric problems in not doing so.

Laurasia probably consisted of North America, Greenland, Eurasia west of the Urals, and Iberia (Spain). It is likely that the northwestern margin of Laurasia lay much further east than present-day coastlines suggest. This is inferred from the present-day cylindrical equidistant maps (maps 1 and 2, section 1) on which we have indicated the areas that have been affected by Paleozoic, Mesozoic and Cenozoic orogenies.

It is unlikely that eastern Eurasia comprised one continent until the end of the Paleozoic. It probably consisted of a number of smaller continents: China, Kolyma, Siberia, Kazakhstan. We have drawn this continent as a single unit because there is no clear-cut support for any of the several possible reconstructions except from sparse paleomagnetic data (Scotese, Bambach, Barton, Van der Voo & Ziegler, 1979). Therefore we have used data from only the Siberian platform to orientate this block. The faunal evidence provides a better guide to the way in which they may be arranged (Ziegler, Scotese, McKerrow, Johnson & Bambach, 1977, 1979). However, we have chosen to keep these maps independent of geological and paleontologic data. Maps prior to 400 Ma show Laurasia separated into two smaller continents – 'western Eurasia', comprising the European platform, southern Britain, eastern Newfoundland and eastern Spitzbergen, and 'Laurentia', comprising North America, Greenland, northern Britain, western Spitzbergen and western Newfoundland.

It is impossible to determine a world reassembly of the continents for any period prior to the formation of Pangea because none of the presumed intervening ocean floor has been preserved, except as tectonic slices (ophiolites) in old orogenic belts. We assume that the Paleozoic orogenic belts, which cut across present-day continents, mark one or more sutures formed by subduction of old ocean floor and subsequent continental collision. Thus the suture by which Gondwanaland and Laurasia west of the Urals were joined together lies within the Appalachian–Hercynian orogenic belt. Similar sutures lie within the Urals, joining Laurasia west of the Urals with Laurasia to the east, and within the Caledonides, joining a continental area to the west with a continental area to the east.

To make a map we treat each continent (Laurasia, Gondwanaland, etc.) separately. The rotations used to produce the reassembly of the continents are the same as those employed in section 1 for the 180–220 Ma reconstructions of Pangea.

Paleomagnetic data from each continent are rotated to the reference fragment (Africa) and a separate mean north pole calculated. Each continent is then projected as a map. The maps are then superimposed; this is the equivalent process to superimposing the mean north paleomagnetic pole of each Paleozoic continent. Keeping Gondwanaland fixed, the other continents are rotated about the north pole, by an arbitrary angle, until the overlap of continental margins is minimised. The final map is then projected as a world map.

The Permo-Triassic problem
The main problem in compiling pre-Triassic maps is that there is no reliable way of determining how long Pangaea existed. We know that it began to break up in the early Jurassic. We also infer that the Appalachian–Hercynian and Uralian belts are late Paleozoic compressional plate margins, resulting from the collision of the southern (Gondwanaland), northern (Laurasia) and eastern (eastern Eurasia) continents. Thus these blocks must have been separated by oceans at least up to the mid Paleozoic.

There are now sufficient paleomagnetic data to suggest a persistent difference between the polar wandering paths of Gondwanaland and Laurasia up to the early Triassic but it is still difficult to determine the exact time when the blocks coalesced to form the Pangea of maps 41–52. Therefore, during the late Paleozoic, Gondwanaland and Laurasia are orientated as two separate blocks. The divergence of the polar wandering paths prior to the Triassic is such that it is impossible to project these two continents as a map without invoking a large longitude offset (to avoid overlap of continental areas). The only escape from this complication would be to invoke failure of the geocentre axial dipole model, which otherwise seems acceptable throughout Paleozoic time. Abandoning this model undermines the entire basis of compiling Paleozoic composites and of orientating Mesozoic and Cenozoic reassemblies. Maintaining the paleomagnetic model, then, implies large longitude relative motion between Gondwanaland and Laurasia in the Permo-Triassic.

Compromise Pangea reconstructions have been proposed (e.g., Irving, 1977), but they are neither based on specific geological arguments, nor do they conform precisely to the observed paleomagnetic data.

In this volume we depict the elimination of the longitude offset as having occurred between 240 and 220 Ma. This may be exaggerated, and the relative motion between Laurasia and Gondwanaland may not have ceased until as late as 200 or 190 Ma.

Although this solution is not unique, it is consistent with displacement having occurred simply by large-scale (transform) shear motion between Laurasia and Gondwanaland. The model does not invoke the creation and destruction of new ocean basins, for which there is no geological evidence, and it is amenable to testing.

Reliability of the maps

The errors associated with these maps are threefold. Firstly there are errors in the reconstruction of each continent. Since we are using the same rotation data as in section 1, the errors are the same. We believe them to be negligible compared with the errors associated with the paleomagnetic data. Secondly there is the dispersion amongst the poles selected for the continents at each age. This error is reflected in the increased size of the alpha-95 radius, compared with the maps of section 1. The poles have been selected using the same reliability criteria, and averaged in the same manner, as in section 1. In general there are fewer data available, and those which exist must be divided into at least three smaller geographic groups — reducing N and increasing alpha-95.

Lastly, because the longitude separation is indeterminate, the relative longitude separations of the continents are arbitrary. The continents can be fixed with respect to one another only by different methods, particularly those

drawing on fossil distributions. The reader is free to rearrange the relative separations of all the continental fragments without in any way conflicting with the data used in their reconstruction. However, latitude changes outside the evaluated errors are not permitted by the data. These maps provide a suitable framework for workers, especially in the fields of numerical and comparative taxonomy.

Because Gondwanaland has been used as a reference continent for the arbitrary longitude rotations, the movement of the grid north pole with respect to Africa in its present-day coordinates is simply the Gondwanaland mean polar wandering path.

The last maps of the series (560 Ma) have very large alpha-95 values associated with them. They are the least well-documented maps of the set and their associated uncertainties are aggravated by the rapidity of apparent polar wandering of some continental fragments around that time. The most obvious effects of these uncertainties on the maps relate to the latitude of Gondwanaland and the orientation of eastern Eurasia. The rotation of eastern Eurasia, depicted as having occurred between 560 and 520 Ma, may well have begun rather earlier, and the amount of rotation between 560 and 520 Ma may therefore have been less than is depicted on the maps.

There are two further aspects of the entire series of Paleozoic maps which require comment. Firstly, the grouping of present-day continents to form the Paleozoic continents is based on an understanding of the geological evidence of late Paleozoic orogeny and possible continental suturing. Only in the case of eastern Eurasia is there substantial evidence that we have incorrectly aggregated separate fragments; we did so both to aid recognition of the whole region and for want of sufficient evidence as to where the separate fragments should otherwise be placed.

The Paleozoic continents have been projected in relative positions such that the progression from one map to the next is along the shortest route. Whilst this is the most likely pattern of motion we must stress that it is not known whether this is the actual pattern. For example, the three main continents may have existed in a different order, as one traverses along a particular latitude circle.

By progression from one map to the next, the formation of the Triassic Pangea (and thence on to the present day) is geometrically possible. This does not, however, prove that alternative orders are impossible, nor is it claimed that the order presented here is uniquely compatible with geological evidence. Because of these complications, the sum of all possible errors in the Paleozoic maps is much

greater than that associated with the Mesozoic maps.
Elimination of ambiguity, refinements of the pattern,
and improvements in precision will depend on future
studies aimed at integrating geological information
with new paleomagnetic data.

Acknowledgements

The maps are the outcome of a project supported by the
Natural Environmental Research Council (GR/3/2277).
The authors thank I. O. Norton and J. G. Sclater for pro-
viding us with a reprint of their paper on the evolution
of the Indian Ocean, whose results are incorporated into
the maps in section 1.

Map 53
240 million years
approx. Tatarian (late Permian)

Cylindrical equidistant
Gondwanaland: $N = 29$ alpha-95 = 5.1
Laurasia: $N = 80$ alpha-95 = 3.5

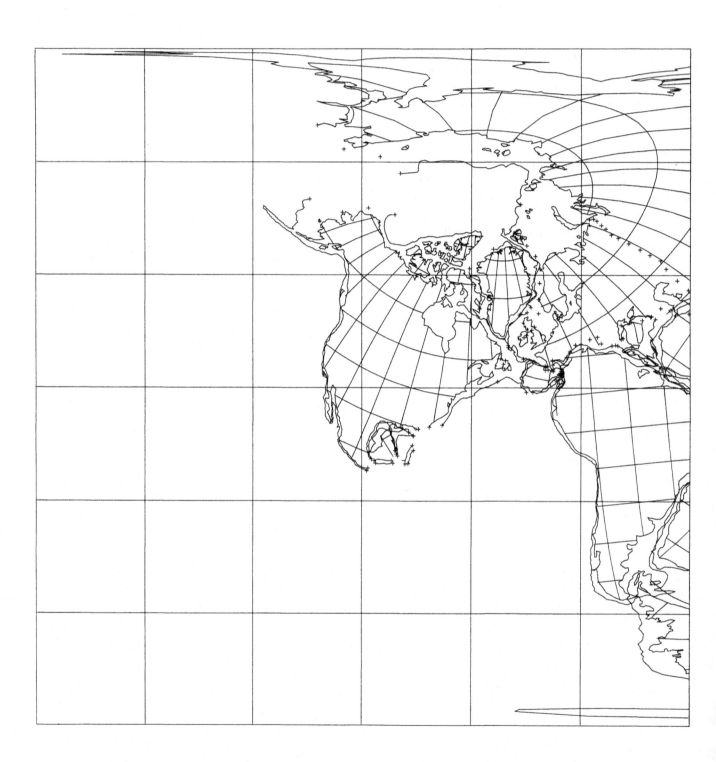

Map 54
240 million years
approx. Tatarian (late Permian)

Cylindrical equidistant
Gondwanaland: $N = 29$ alpha-95 = 5.1
Laurasia: $N = 80$ alpha-95 = 3.5

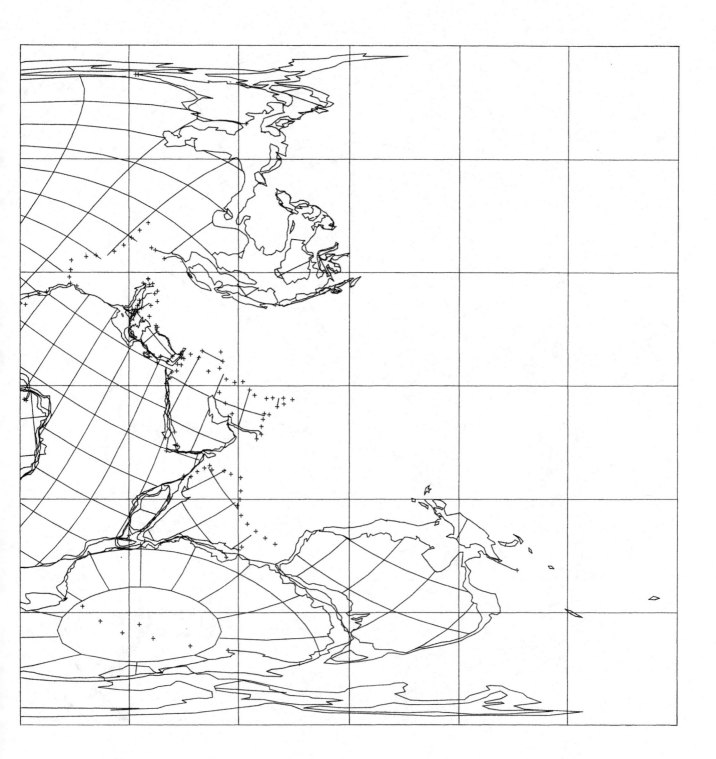

Map 55
240 million years
approx. Tatarian (late Permian)

North polar Lambert equal-area
Gondwanaland: $N = 29$ alpha-95 = 5.1
Laurasia: $N = 80$ alpha-95 = 3.5

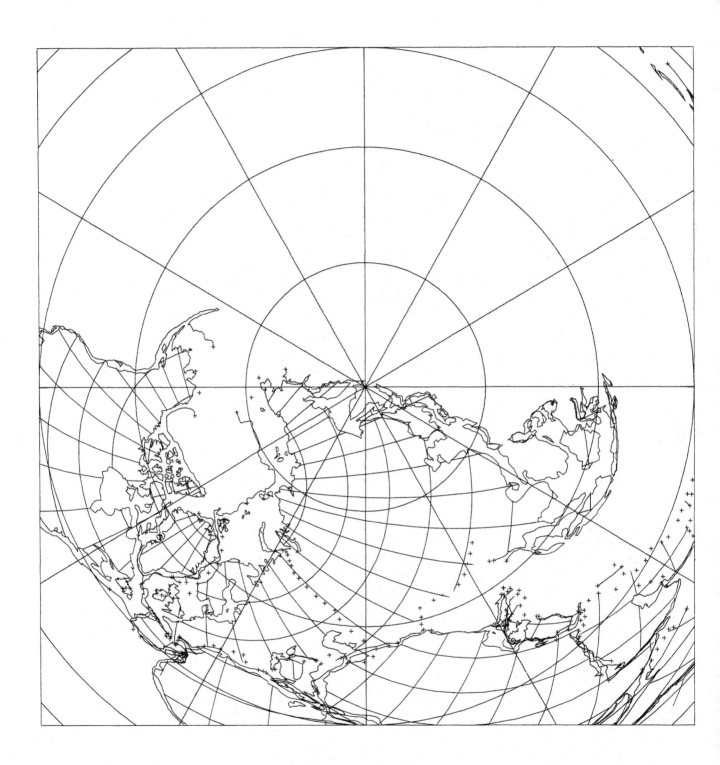

Map 56
240 million years
approx. Tatarian (late Permian)

South polar Lambert equal-area
Gondwanaland: $N = 29$ alpha-95 = 5.1
Laurasia: $N = 80$ alpha-95 = 3.5

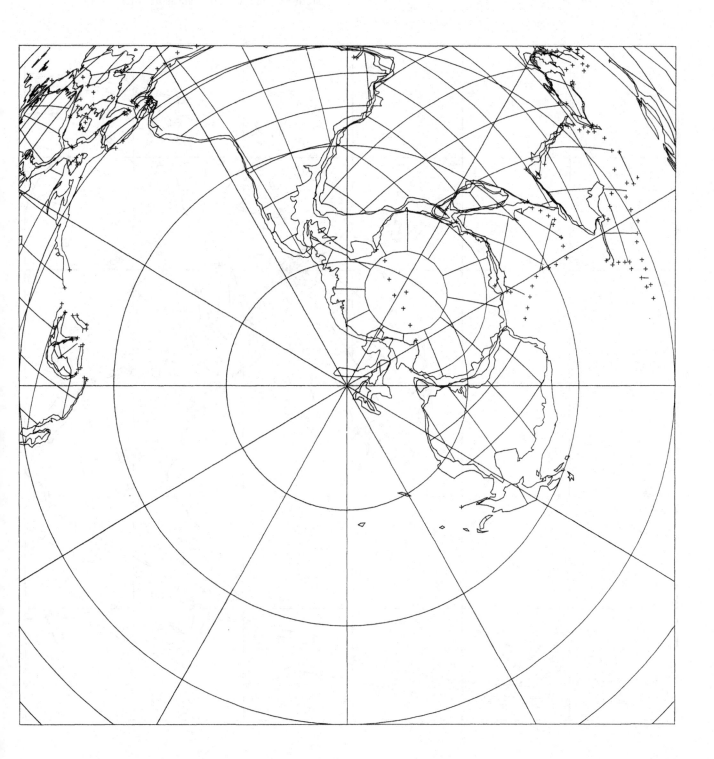

Map 57
280 million years
approx. Sakmarian (very early Permian)

Cylindrical equidistant
Gondwanaland: $N = 27$ alpha-95 = 6.1
Laurasia: $N = 74$ alpha-95 = 1.9
eastern Eurasia: $N = 10$ alpha-95 = 12.7

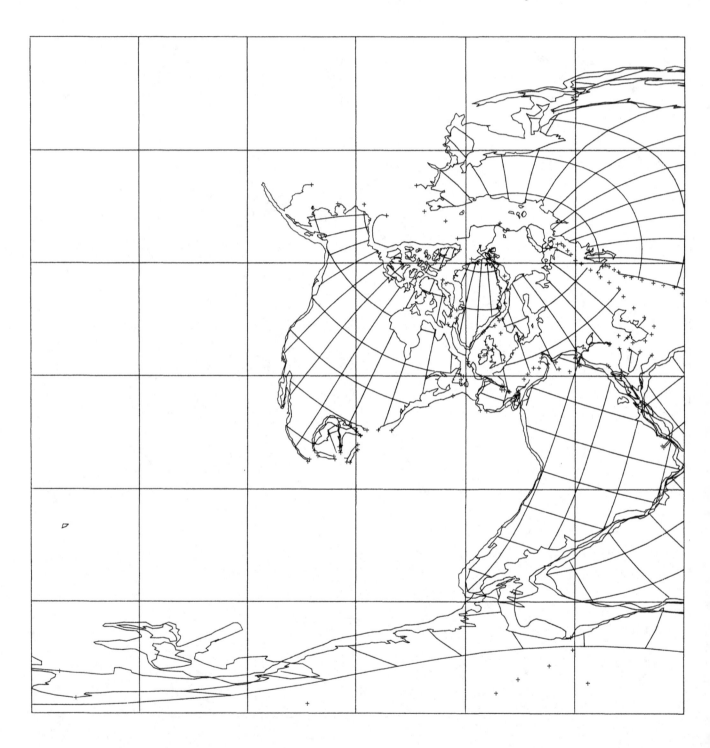

Map 58

280 million years
approx. Sakmarian (very early Permian)

Cylindrical equidistant
Gondwanaland: $N = 27$ alpha-95 = 6.1
Laurasia: $N = 74$ alpha-95 = 1.9
eastern Eurasia: $N = 10$ alpha-95 = 12.7

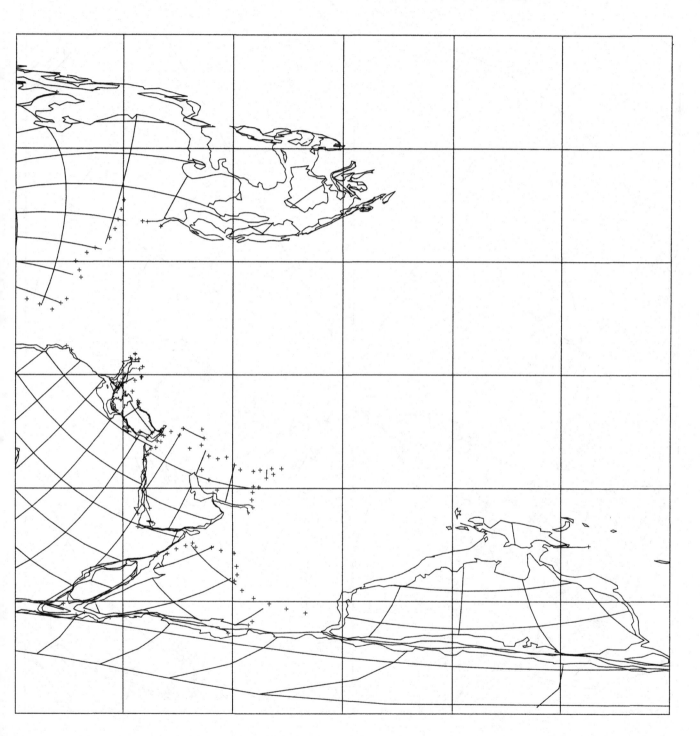

Map 59
280 million years
approx. Sakmarian (very early Permian)

North polar Lambert equal-area
Gondwanaland: $N = 27$ alpha-95 = 6.1
Laurasia: $N = 74$ alpha-95 = 1.9
eastern Eurasia: $N = 10$ alpha-95 = 12.7

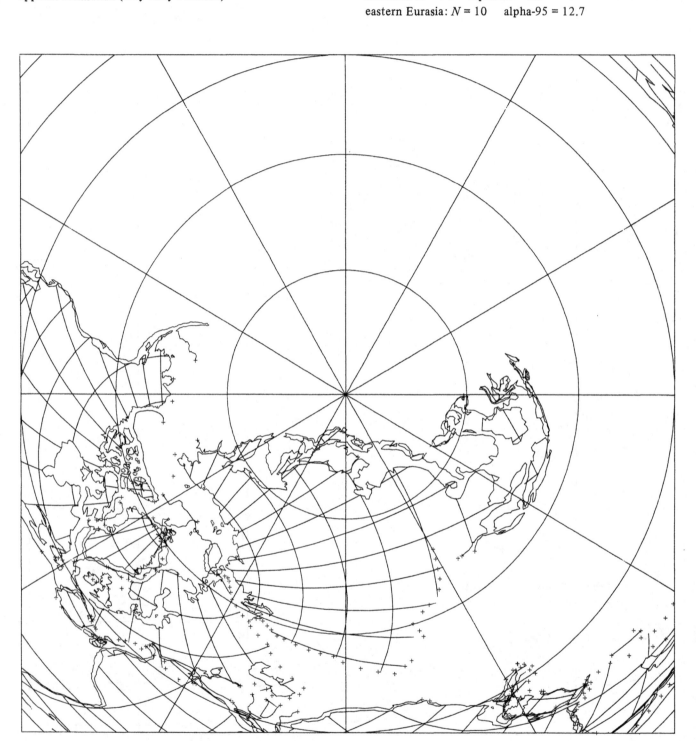

Map 60
280 million years
approx. Sakmarian (very early Permian)

South polar Lambert equal-area
Gondwanaland: $N = 27$ alpha-95 = 6.1
Laurasia: $N = 74$ alpha-95 = 1.9
eastern Eurasia: $N = 10$ alpha-95 = 12.7

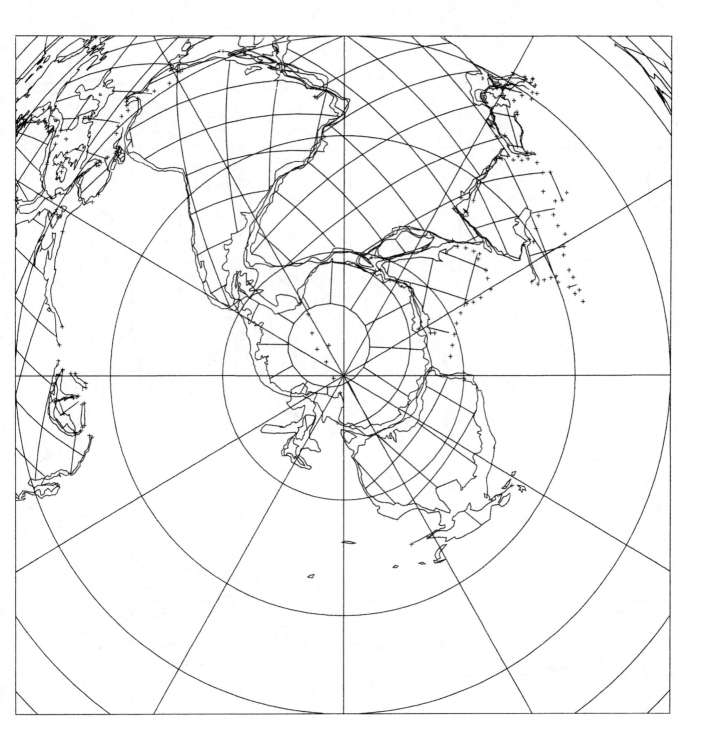

Map 61
320 million years
approx. Namurian (mid Carboniferous)

Cylindrical equidistant
Gondwanaland: $N = 21$ alpha-95 = 8.3
Laurasia: $N = 76$ alpha-95 = 2.9
eastern Eurasia: $N = 11$ alpha-95 = 12.8

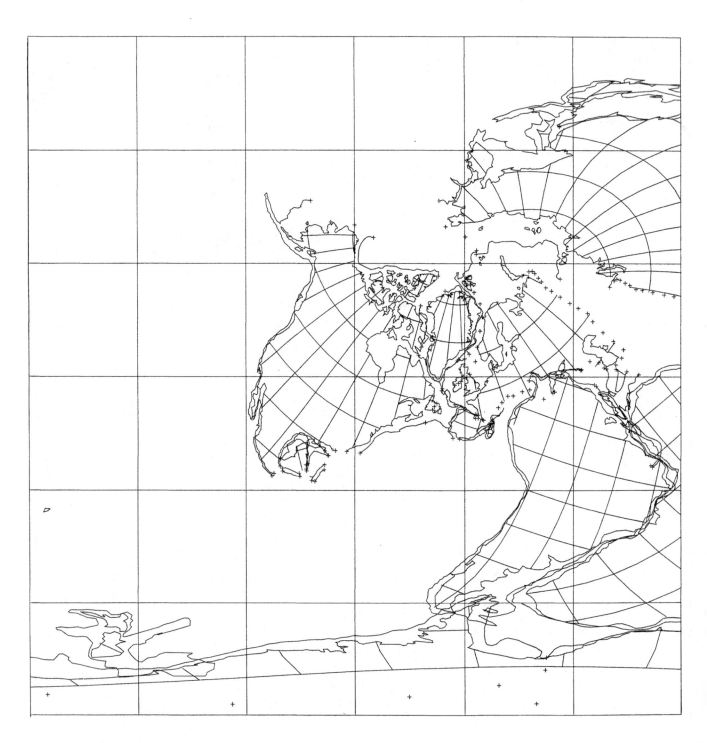

Map 62
320 million years
approx. Namurian (mid Carboniferous)

Cylindrical equidistant
Gondwanaland: $N = 21$ alpha-95 = 8.3
Laurasia: $N = 76$ alpha-95 = 2.9
eastern Eurasia: $N = 11$ alpha-95 = 12.8

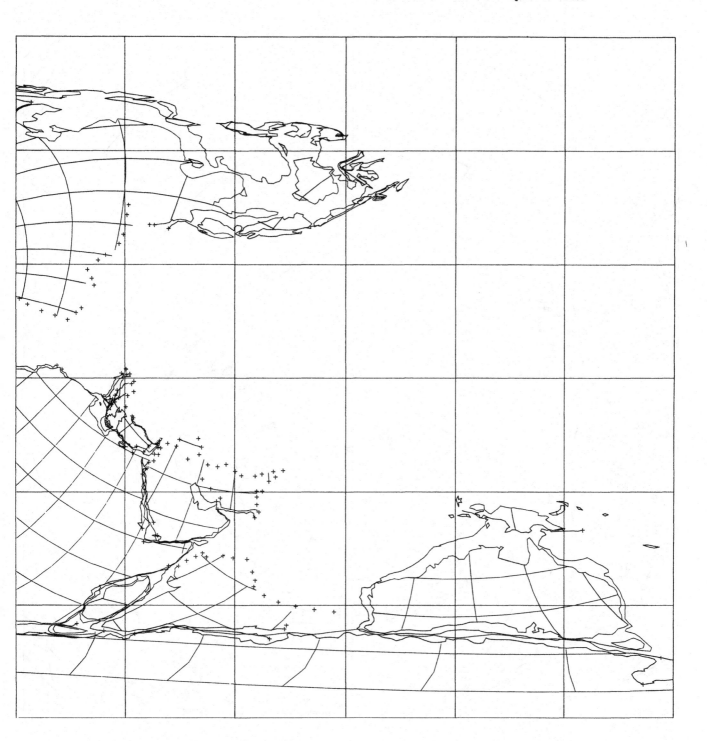

Map 63
320 million years
approx. Namurian (mid Carboniferous)

North polar Lambert equal-area
Gondwanaland: $N = 21$ alpha-95 = 8.3
Laurasia: $N = 76$ alpha-95 = 2.9
eastern Eurasia: $N = 11$ alpha-95 = 12.8

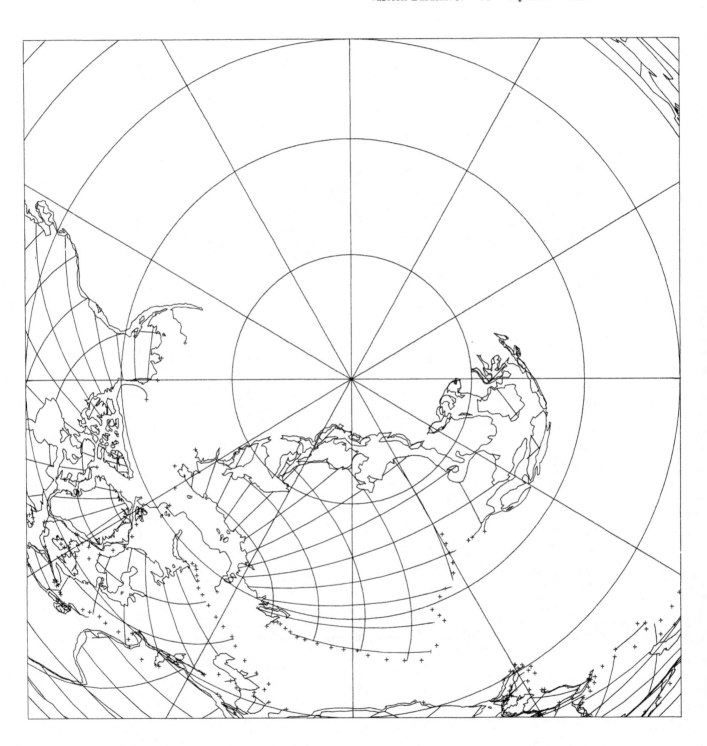

Map 64
320 million years
approx. Namurian (mid Carboniferous)

South polar Lambert equal-area
Gondwanaland: $N = 21$ alpha-95 = 8.3
Laurasia: $N = 76$ alpha-95 = 2.9
eastern Eurasia: $N = 11$ alpha-95 = 12.8

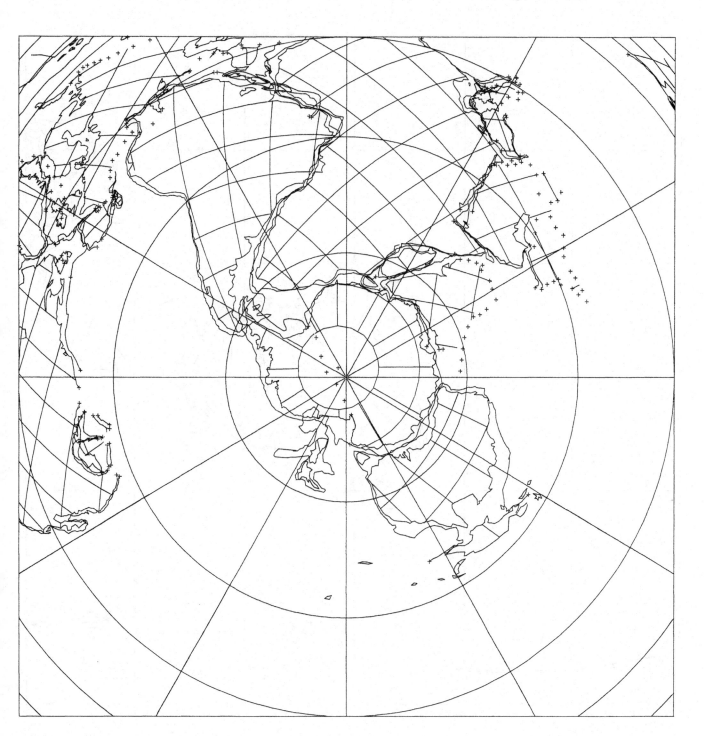

Map 65
360 million years
approx. Frasnian (late Devonian)

Cylindrical equidistant
Gondwanaland: $N = 7$ alpha-95 = 19.8
Laurasia: $N = 33$ alpha-95 = 6.7
eastern Eurasia: $N = 9$ alpha-95 = 9.7

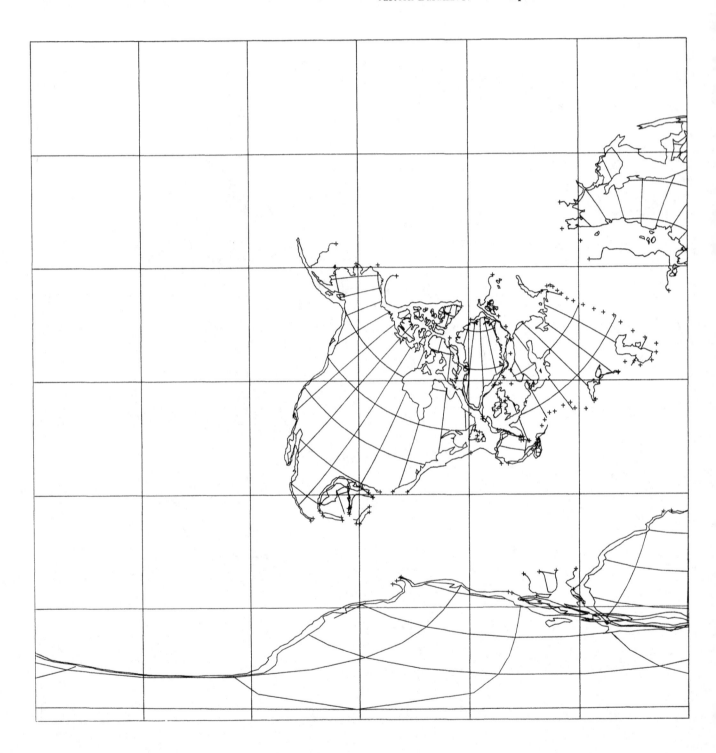

Map 66
360 million years
approx. Frasnian (late Devonian)

Cylindrical equidistant
Gondwanaland: $N = 7$ alpha-95 = 19.8
Laurasia: $N = 33$ alpha-95 = 6.7
eastern Eurasia: $N = 9$ alpha-95 = 9.7

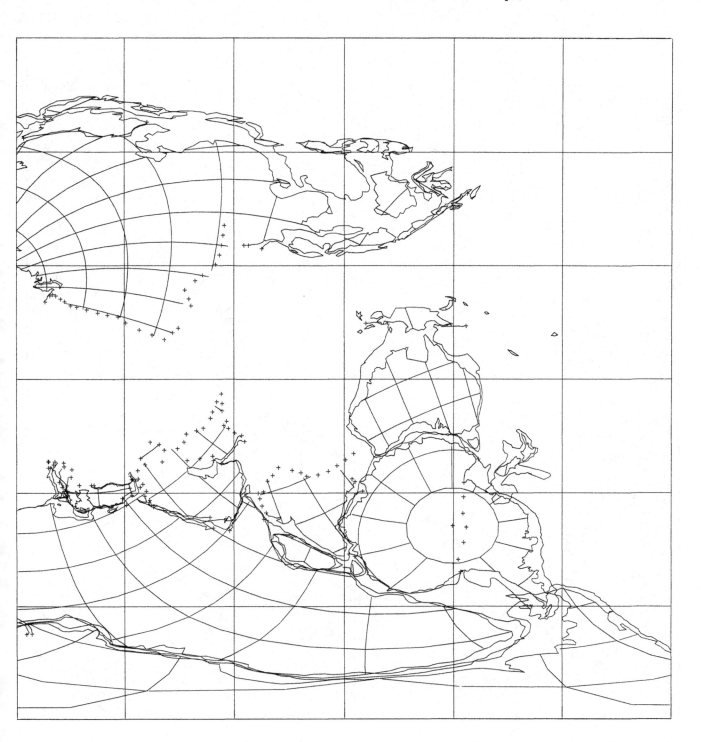

Map 67
360 million years
approx. Frasnian (late Devonian)

North polar Lambert equal-area
Gondwanaland: $N = 7$ alpha-95 = 19.8
Laurasia: $N = 33$ alpha-95 = 6.7
eastern Eurasia: $N = 9$ alpha-95 = 9.7

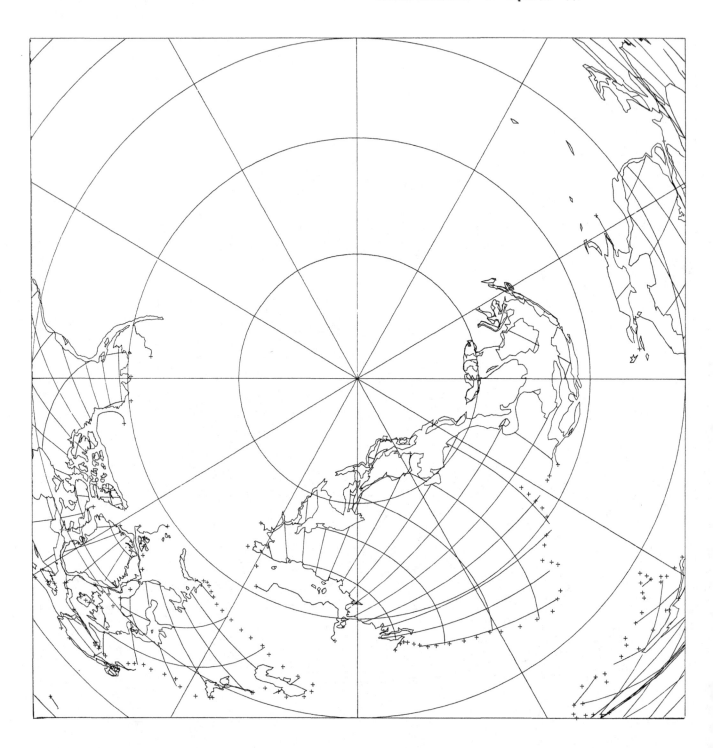

Map 68
360 million years
approx. Frasnian (late Devonian)

South polar Lambert equal-area
Gondwanaland: $N = 7$ alpha-95 = 19.8
Laurasia: $N = 33$ alpha-95 = 6.7
eastern Eurasia: $N = 9$ alpha-95 = 9.7

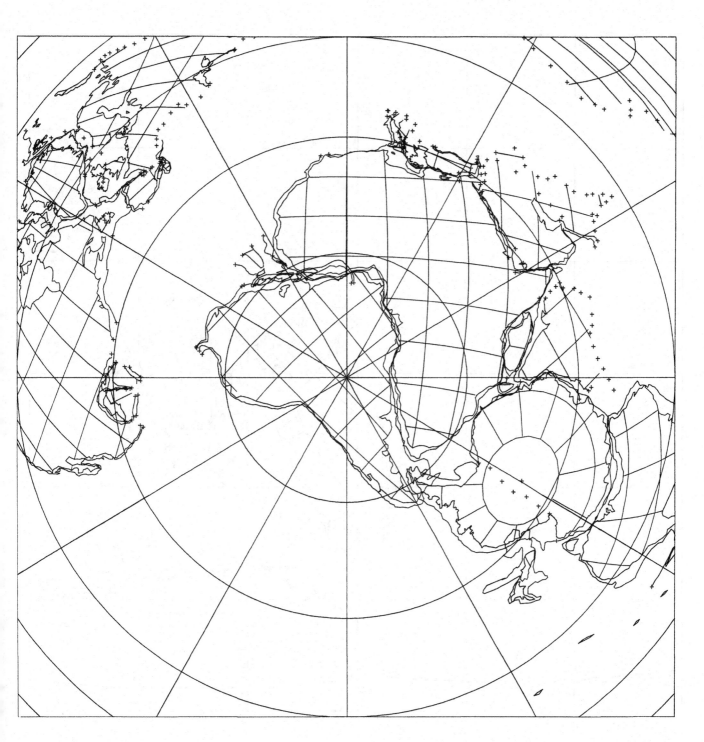

Map 69
400 million years
Ludlovian (late Silurian)

Cylindrical equidistant
Gondwanaland: $N = 5$ alpha-95 = 29.4
Laurasia: $N = 27$ alpha-95 = 10.4
eastern Eurasia: $N = 5$ alpha-95 = 19.2

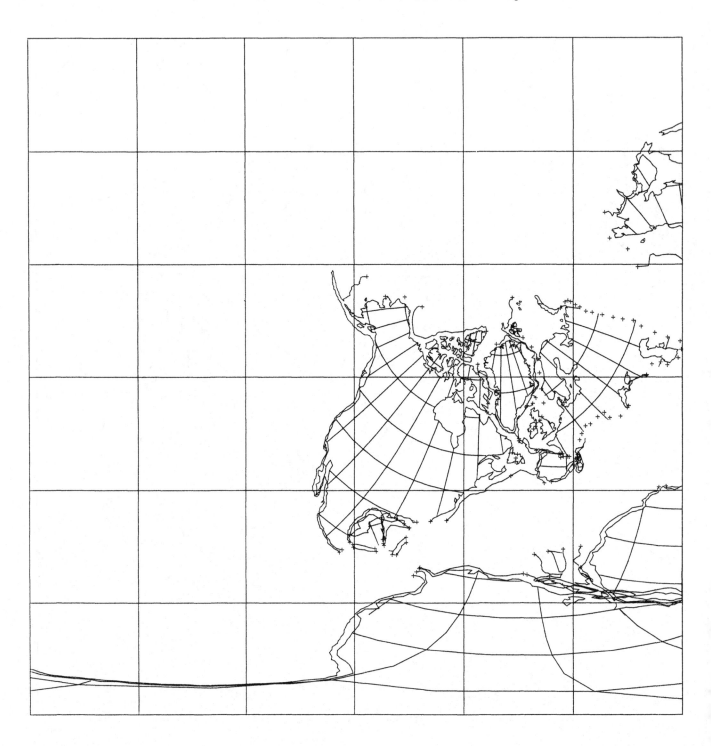

Map 70
400 million years
Ludlovian (late Silurian)

Cylindrical equidistant
Gondwanaland: $N = 5$ alpha-95 = 29.4
Laurasia: $N = 27$ alpha-95 = 10.4
eastern Eurasia: $N = 5$ alpha-95 = 19.2

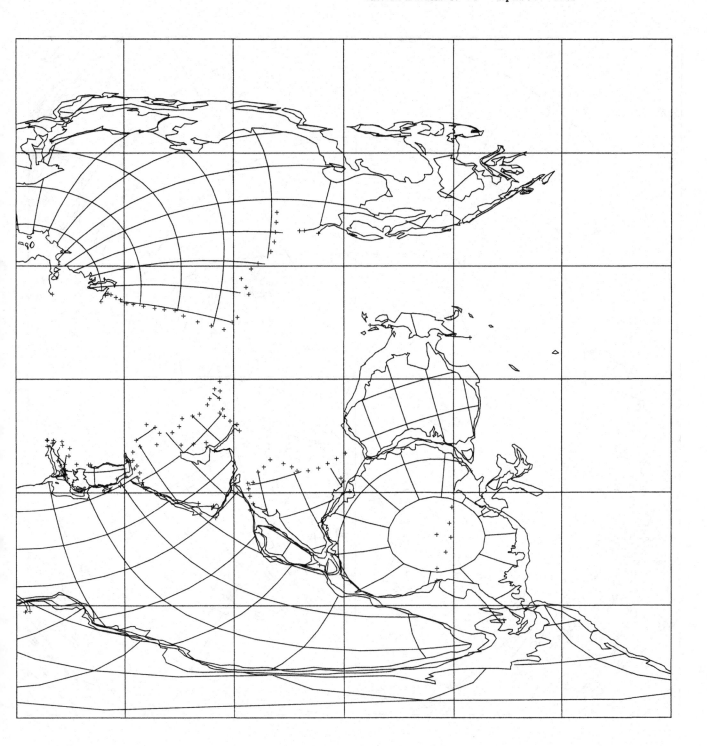

Map 71
400 million years
Ludlovian (late Silurian)

North polar Lambert equal-area
Gondwanaland: $N = 5$ alpha-95 = 29.4
Laurasia: $N = 27$ alpha-95 = 10.4
eastern Eurasia: $N = 5$ alpha-95 = 19.2

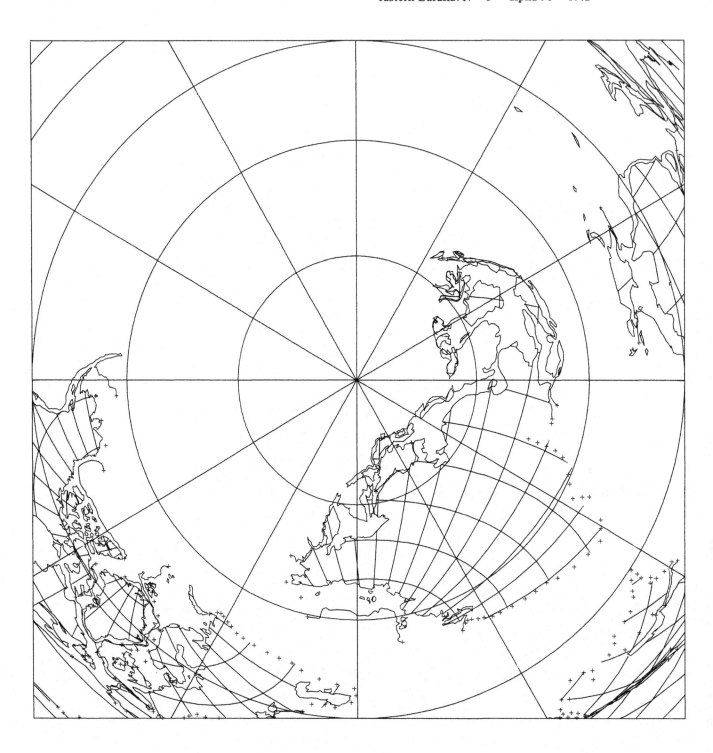

Map 72
400 million years
Ludlovian (late Silurian)

South polar Lambert equal-area
Gondwanaland: $N = 5$ alpha-95 = 29.4
Laurasia: $N = 27$ alpha-95 = 10.4
eastern Eurasia: $N = 5$ alpha-95 = 19.2

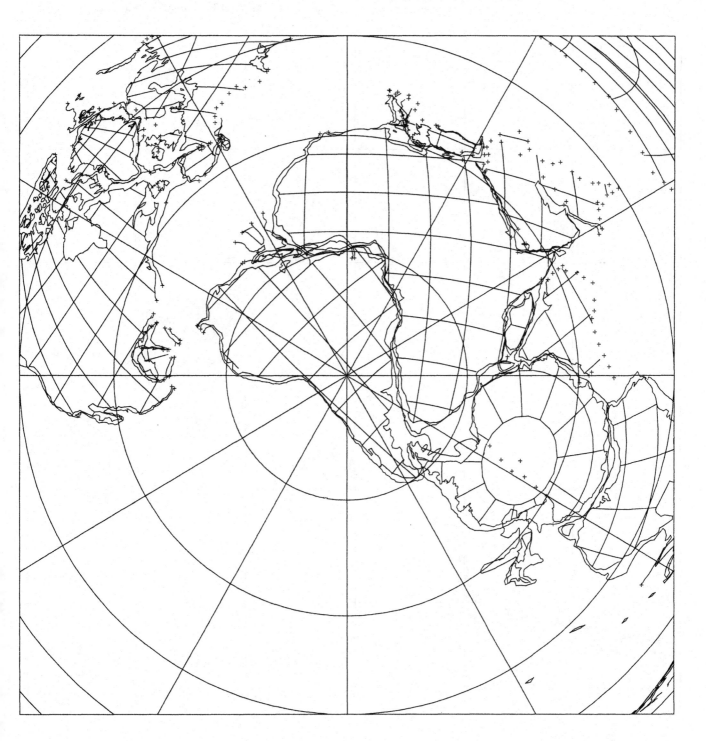

Map 73
440 million years
approx. Ashgillian (late Ordovician)

Cylindrical equidistant
Gondwanaland: $N = 12$ alpha-95 = 18.2
Laurentia: $N = 14$ alpha-95 = 16.1
western Eurasia: $N = 23$ alpha-95 = 19.8
eastern Eurasia: $N = 17$ alpha-95 = 8.6

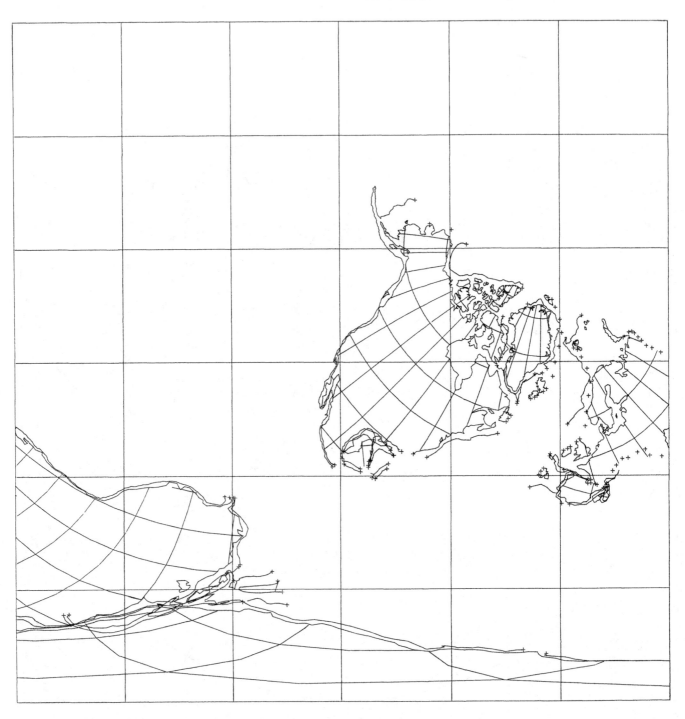

Map 74
440 million years
approx. Ashgillian (late Ordovician)

Cylindrical equidistant
Gondwanaland: $N = 12$ alpha-95 = 18.2
Laurentia: $N = 14$ alpha-95 = 16.1
western Eurasia: $N = 23$ alpha-95 = 19.8
eastern Eurasia: $N = 17$ alpha-95 = 8.6

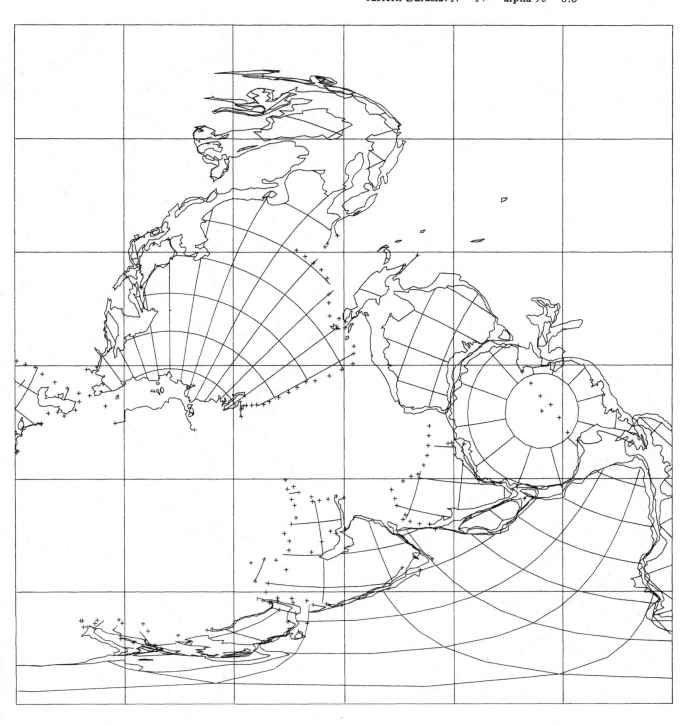

Map 75
440 million years
approx. Ashgillian (late Ordovician)

North polar Lambert equal-area
Gondwanaland: $N = 12$ alpha-95 = 18.2
Laurentia: $N = 14$ alpha-95 = 16.1
western Eurasia: $N = 23$ alpha-95 = 19.8
eastern Eurasia: $N = 17$ alpha-95 = 8.6

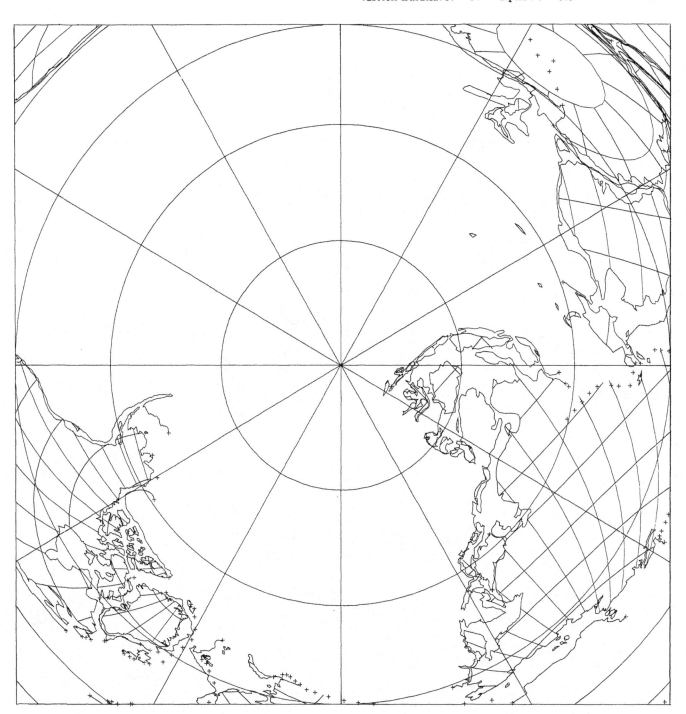

Map 76
440 million years
approx. Ashgillian (late Ordovician)

South polar Lambert equal-area
Gondwanaland: $N = 12$ alpha-95 = 18.2
Laurentia: $N = 14$ alpha-95 = 16.1
western Eurasia: $N = 23$ alpha-95 = 19.8
eastern Eurasia: $N = 17$ alpha-95 = 8.6

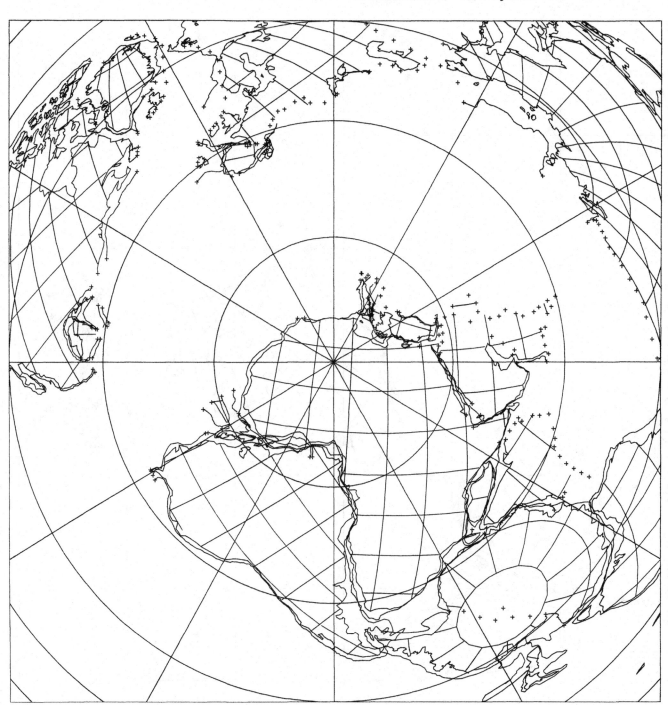

85

Map 77
480 million years
approx. Arenigian (early Ordovician)

Cylindrical equidistant
Gondwanaland: $N = 12$ alpha-95 = 14.1
Laurentia: $N = 10$ alpha-95 = 19.8
western Eurasia: $N = 12$ alpha-95 = 19.8
eastern Eurasia: $N = 12$ alpha-95 = 6.2

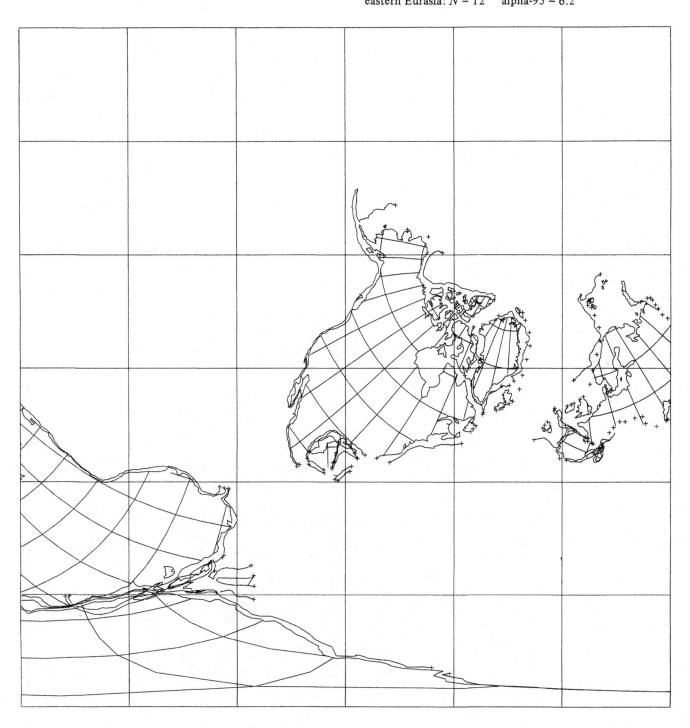

Map 78
480 million years
approx. Arenigian (early Ordovician)

Cylindrical equidistant
Gondwanaland: $N = 12$ alpha-95 = 14.1
Laurentia: $N = 10$ alpha-95 = 19.8
western Eurasia: $N = 12$ alpha-95 = 19.8
eastern Eurasia: $N = 12$ alpha-95 = 6.2

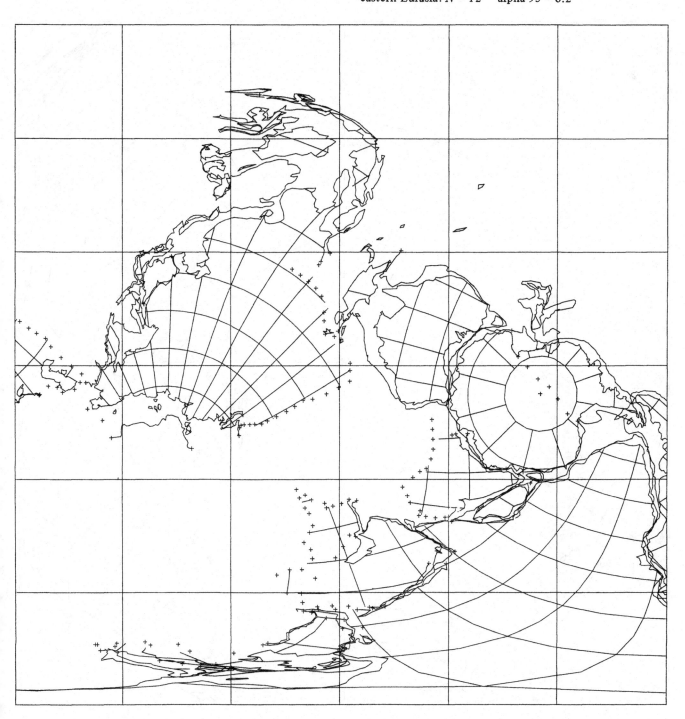

Map 79
480 million years
approx. Arenigian (early Orovician)

North polar Lambert equal-area
Gondwanaland: $N = 12$ alpha-95 = 14.1
Laurentia: $N = 10$ alpha-95 = 19.8
western Eurasia: $N = 12$ alpha-95 = 19.8
eastern Eurasia: $N = 12$ alpha-95 = 6.2

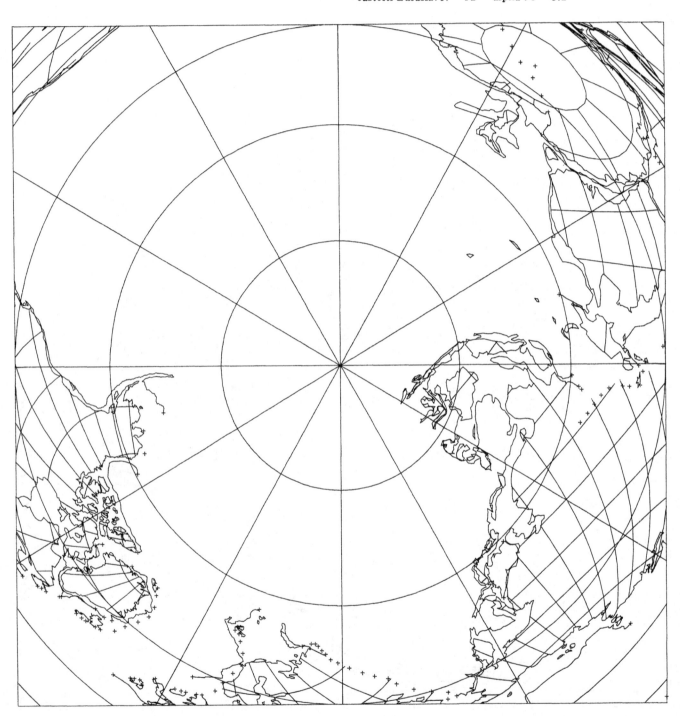

Map 80
480 million years
approx. Arenigian (early Ordovician)

South polar Lambert equal-area
Gondwanaland: $N = 12$ alpha-95 = 14.1
Laurentia: $N = 10$ alpha-95 = 19.8
western Eurasia: $N = 12$ alpha-95 = 19.8
eastern Eurasia: $N = 12$ alpha-95 = 6.2

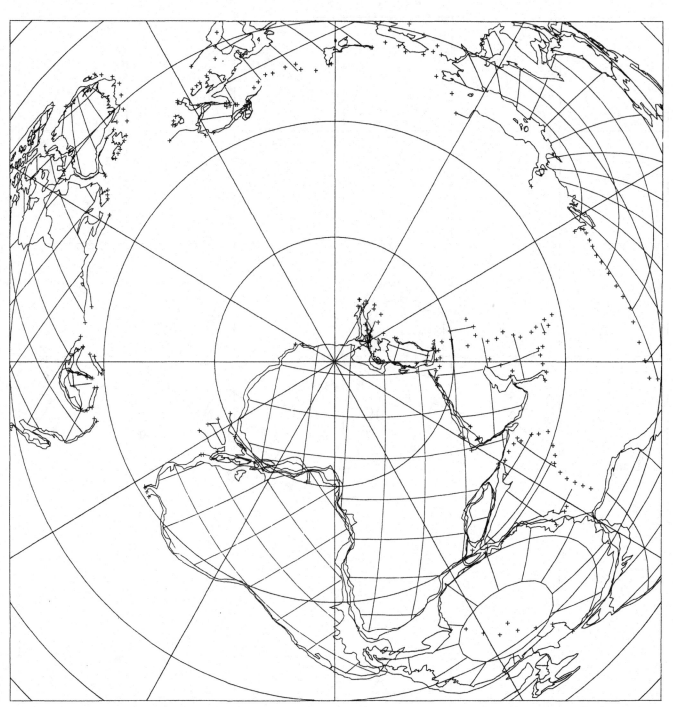

Map 81
520 million years
approx. Mayan (mid Cambrian)

Cylindrical equidistant
Gondwanaland: $N = 13$ alpha-95 = 19.8
Laurentia: $N = 9$ alpha-95 = 25.4
western Eurasia: $N = 5$ alpha-95 = 30.7
eastern Eurasia: $N = 15$ alpha-95 = 4.1

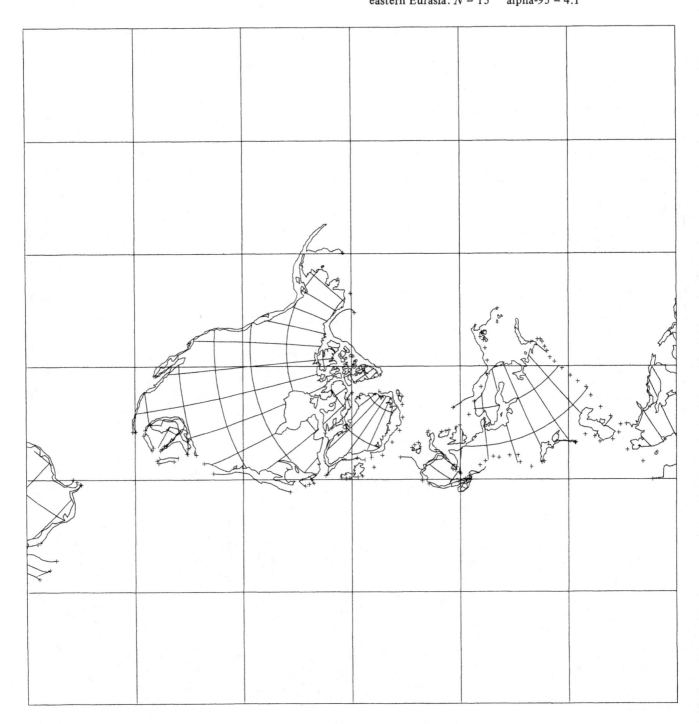

Map 82
520 million years
approx. Mayan (mid Cambrian)

Cylindrical equidistant
Gondwanaland: $N = 13$ alpha-95 = 19.8
Laurentia: $N = 9$ alpha-95 = 25.4
western Eurasia: $N = 5$ alpha-95 = 30.7
eastern Eurasia: $N = 15$ alpha-95 = 4.1

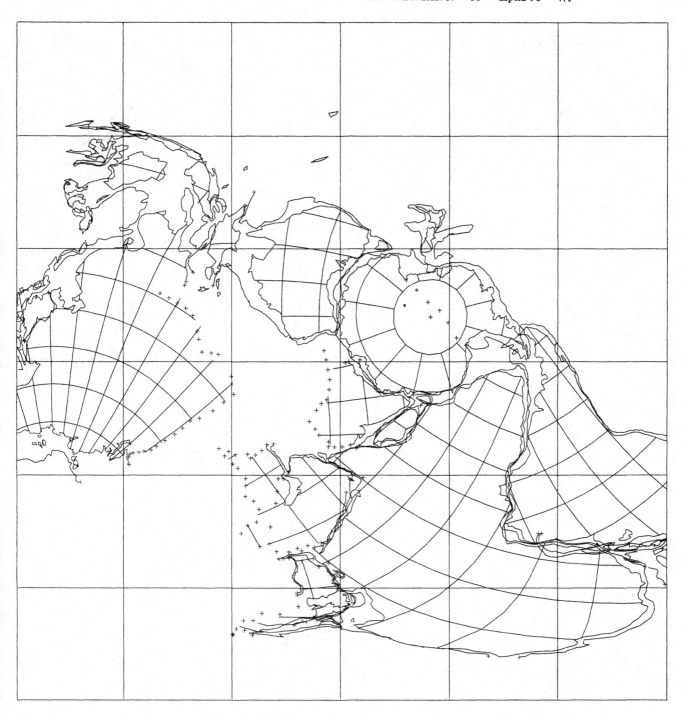

Map 83
520 million years
approx. Mayan (mid Cambrian)

North polar Lambert equal-area
Gondwanaland: $N = 13$ alpha-95 = 19.8
Laurentia: $N = 9$ alpha-95 = 25.4
western Eurasia: $N = 5$ alpha-95 = 30.7
eastern Eurasia: $N = 15$ alpha-95 = 4.1

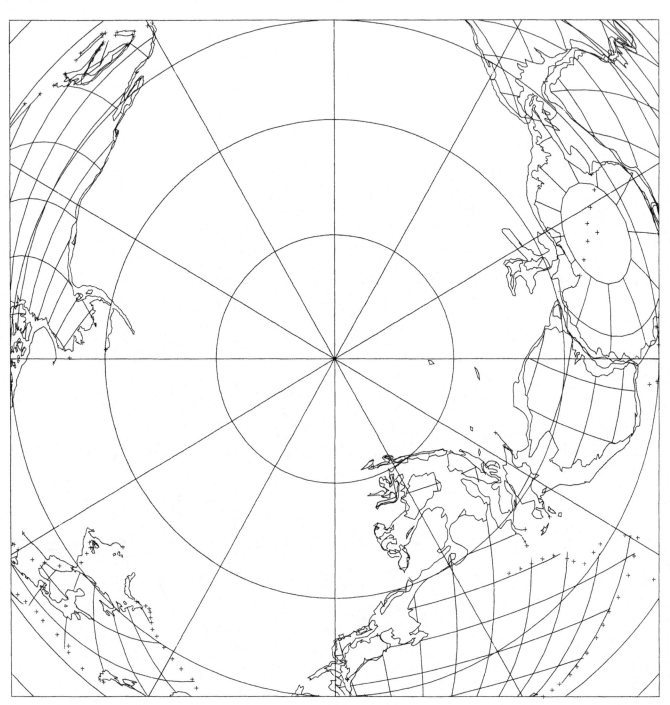

Map 84
520 million years
approx. Mayan (mid Cambrian)

South polar Lambert equal-area
Gondwanaland: $N = 13$ alpha-95 = 19.8
Laurentia: $N = 9$ alpha-95 = 25.4
western Eurasia: $N = 5$ alpha-95 = 30.7
eastern Eurasia: $N = 15$ alpha-95 = 4.1

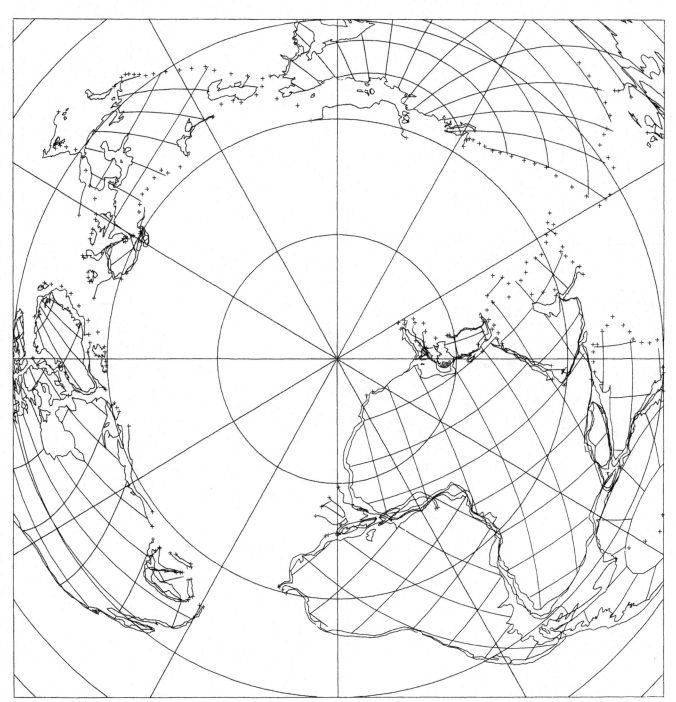

Map 85
560 million years
Aldanian (early Cambrian)

Cylindrical equidistant
Gondwanaland: $N = 11$ alpha-95 = 32.8
Laurentia: $N = 9$ alpha-95 = 30.6
western Eurasia: $N = 5$ alpha-95 = 20.5
eastern Eurasia: $N = 5$ alpha-95 = 21.3

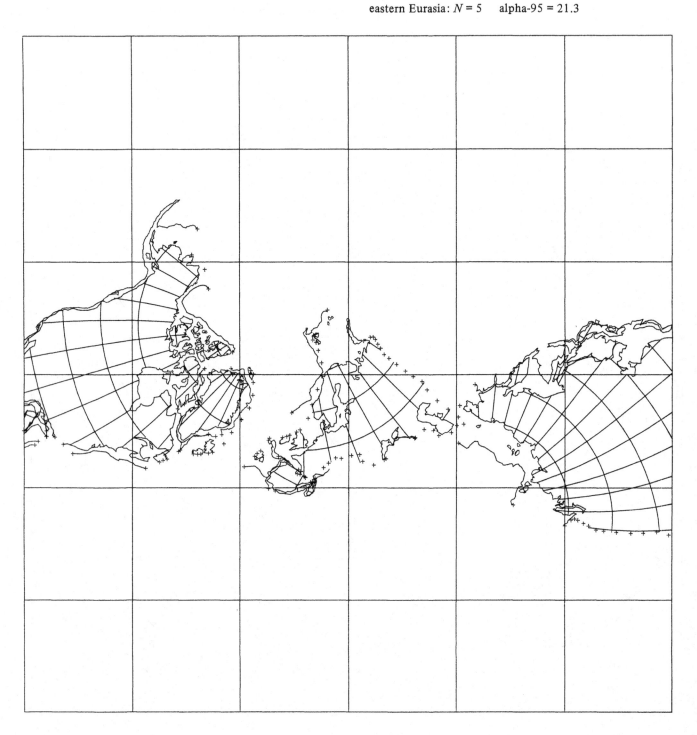

94

Map 86
560 million years
Aldanian (early Cambrian)

Cylindrical equidistant
Gondwanaland: $N = 11$ alpha-95 = 32.8
Laurentia: $N = 9$ alpha-95 = 30.6
western Eurasia: $N = 5$ alpha-95 = 20.5
eastern Eurasia: $N = 5$ alpha-95 = 21.3

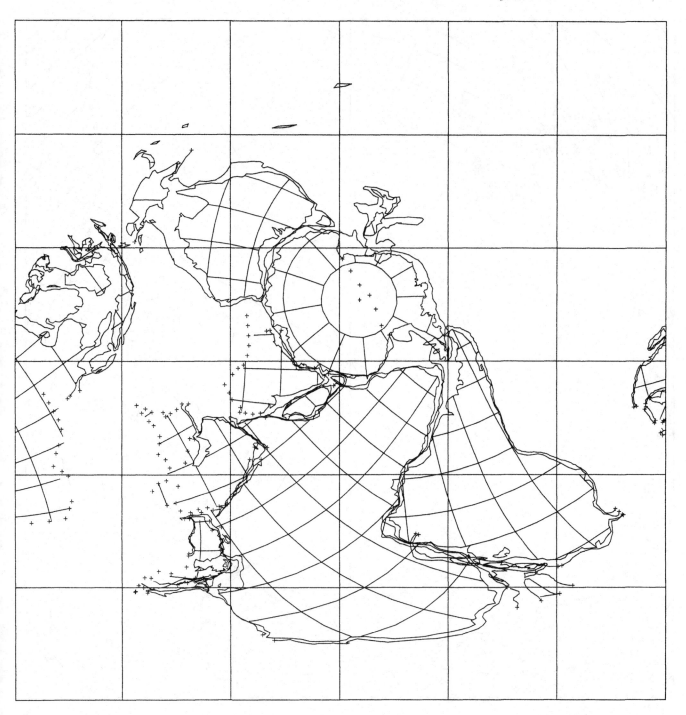

Map 87
560 million years
Aldanian (early Cambrian)

North polar Lambert equal-area
Gondwanaland: $N = 11$ alpha-95 = 32.8
Laruentia: $N = 9$ alpha-95 = 30.6
western Eurasia: $N = 5$ alpha-95 = 20.5
eastern Eurasia: $N = 5$ alpha-95 = 21.3

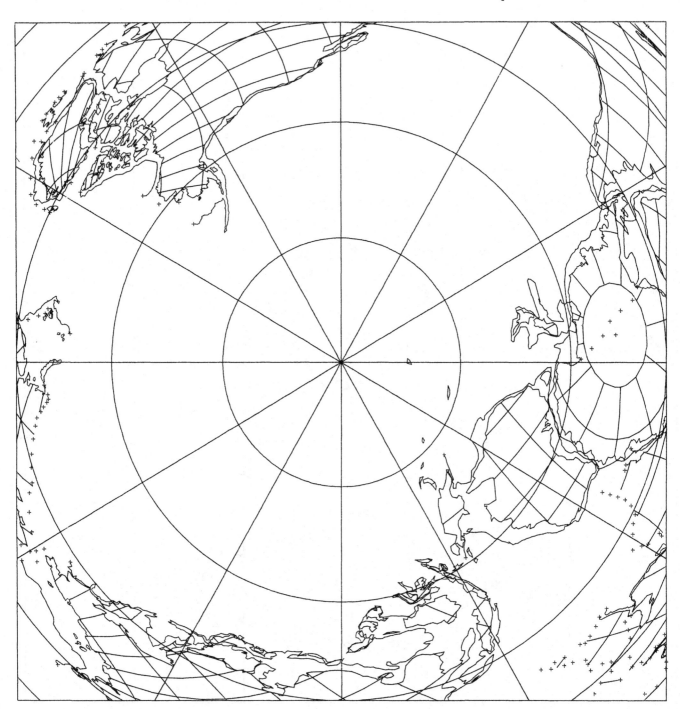

Map 88
560 million years
Aldanian (early Cambrian)

South polar Lambert equal-area
Gondwanaland: $N = 11$ alpha-95 = 32.8
Laurentia: $N = 9$ alpha-95 = 30.6
western Eurasia: $N = 5$ alpha-95 = 20.5
eastern Eurasia: $N = 5$ alpha-95 = 21.3

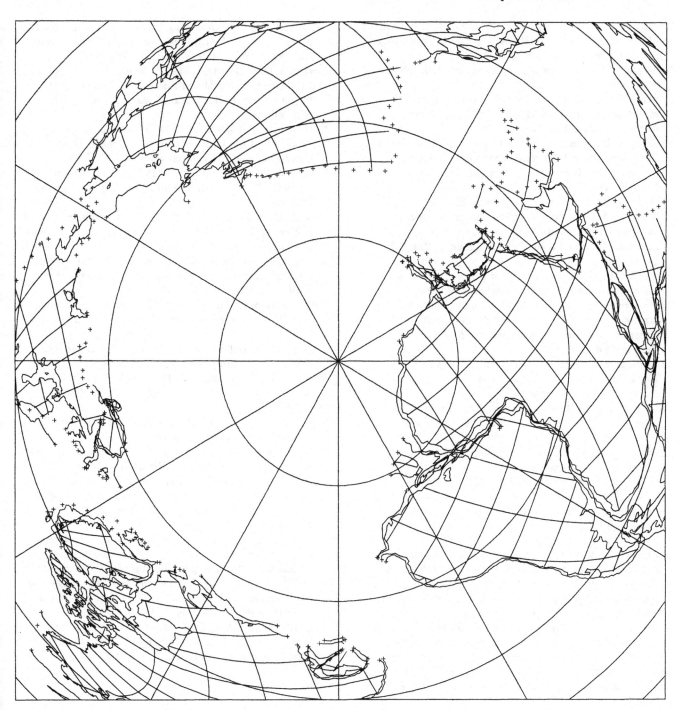

Bibliography

Bullard, E. C., Everett, J. E. & Smith, A. G. (1965). The fit of the continents around the Atlantic. *Phil. Trans. Roy. Soc., London, Ser. A*, **258**, 41–51.

Fisher, R. A. (1953). Dispersion on a sphere. *Proc. Roy. Soc., London, Ser. A*, **217**, 295–305.

Freeland, G. L. & Dietz, R. S. (1971). Plate tectonic evolution of Caribbean and Gulf of Mexico Region. *Nature, London*, **232**, 20–3.

Harland, W. B., Smith, A. G. & Wilcock, B., (1964), (eds.). *The Phanerozoic Time-scale.* (Symposium dedicated to Professor Arthur Holmes.) *Q. J. Geol. Soc., London*, **120S**, 1–458.

Heirtzler, J. R., Dickson, G. O., Herron, E. F. M., Pitman, W. C., III & Le Pichon, X. (1968). Marine magnetic anomalies, geomagnetic field reversals and motions of the ocean floor and continents. *J. Geophys. Res.*, **73**, 2119–36.

Irving, E. (1977). Drift of the major continental blocks since the Devonian. *Nature, London*, **270**, 304–9.

Kristoffersen, Y. & Talwani, M. (1977). Extinct triple junction south of Greenland and Tertiary motion of Greenland relative to North America. *Bull. Geol. Soc. America*, **88**, 897–907.

La Brecque, J. L., Kent, D. V. & Cande, S. C. (1977). Revised magnetic polarity timescale for the late Cretaceous and Cenozoic time. *Geology*, **5**, 330–5.

Ladd, J. W. (1976). Relative motion of South America with respect to North America and Caribbean tectonics. *Bull. Geol. Soc. America*, **87**, 969–76.

Le Pichon, X., Sibuet, J. C. & Francheteau, F. (1977). The fit of the continents around the North Atlantic Ocean. *Tectonophysics*, **38**, 169–209.

McElhinny, M. W. (1972). Notes on progress in geophysics. Paleomagnetic directions and pole positions. 13. Pole numbers 13/1 to 13/94. *Geophys. J. Roy. Astron. Soc.*, **30**, 281–93.

McElhinny, M. W. (1973). *Paleomagnetism and plate tectonics.* Cambridge: Cambridge University Press.

McElhinny, M. W. & Cowley, J. A. (1977a). Paleomagnetic directions and pole positions. 14. Pole numbers 14/1 to 14/574. *Geophys. J. Roy. Astron. Soc.*, **49**, 313–56.

McElhinny, M. W. & Cowley, J. A. (1977b). Paleomagnetic directions and pole positions. 15. Pole numbers 15/1 to 15/232. *Geophys. J. Roy. Astron. Soc.*, **52**, 259–76.

McKenzie, D. P., Molnar, P. & Davies, D. (1970). Plate tectonics of the Red Sea and east Africa. *Nature, London*, **226**, 243–8.

McKenzie, D. P. & Sclater, J. G. (1971). The evolution of the Indian Ocean since the Late Cretaceous. *Geophys. J. Roy. Astron. Soc.*, **25**, 437–528.

Molnar, P., Atwater, T., Mammerickx, J. & Smith, S. M. (1975). Magnetic anomalies, bathymetry and the tectonic evolution of the South Pacific since the late Cretaceous. *Geophys. J. Roy. Astron. Soc.*, **40**, 383–420.

Norton, I. O. & Molnar, P. (1977). Implications of a revised fit between Australia and Antarctica for the evolution of the Eastern Indian Ocean. *Nature, London*, **267**, 338–40.

Norton, I. O. & Sclater, J. G. (1979). A model for the evolution of the Indian Ocean and the breakup of Gondwanaland. *J. Geophys. Res.*, **84**, 6831–9.

Pitman, W. C., III & Ialwani, M. (1972). Sea floor spreading in the North Atlantic. *Bull. Geol. Soc. America*, **83**, 619–46.

Scotese, C. R., Bambach, R. K., Barton, C., Van der Voo, R. & Ziegler, A. M. (1979) Paleozoic base maps. *J. Geology*, **87**, 217–77.

Smith, A. G. (1971). Alpine deformation and the oceanic areas of the Tethys, Mediterranean and Atlantic. *Bull. Geol. Soc. America*, **82**, 2039–70.

Smith, A. G. & Briden, J. C. (1977). *Mesozoic and Cenozoic paleocontinental maps.* Cambridge: Cambridge University Press.

Smith, A. G. & Hallam, A. (1970). The fit of the southern continents. *Nature, London*, **225**, 139–44.

Talwani, M. & Eldholm, O. (1977). Evolution of the Norwegian–Greenland sea. *Bull. Geol. Soc. America*, **88**, 969–99.

Van Eysinga, F. W. B. (1975). *Geological Time Table.* Amsterdam: Elsevier Scientific.

Weissel, J. K. & Hayes, D. E. (1977). Evolution of the Tasman Sea reappraised. *Earth Planet. Sci. Lett.*, **36**, 77–84.

Weissel, J. K., Hayes, D. E. & Herron, E. M. (1977). Plate tectonic synthesis: the displacements between Australia, New Zealand and Antarctica since the Late Cretaceous. *Mar. Geol.*, **25**, 231–77.

Ziegler, A. M., Scotese, C. R., McKerrow, W. S., Johnson, M. E. & Bambach, R. K. (1977). Paleozoic biogeography of continents bordering the Iapetus (pre Caledonian) and Rheic (pre Hercynian) Oceans. In *Paleontology and Plate tectonics*, ed. R. M. West, *Spec. Publ. Biol. Geol., Milwaukee Public Museum*, **2**, 1–22.

Ziegler, A. M., Scotese, C. R., McKerrow, W. S., Johnson, M. E. & Bambach, R. K. (1979). Paleozoic paleogeography *Ann. Rev. Earth Planet. Sci.*, **7**, 473–502.